Mediation AND THE
Communication Matrix

"*Mediation and the Communication Matrix* is a milestone in communication and media studies, as well as screen studies. C. Kaha Waite integrates media ecology scholarship with phenomenological method as she examines the role that communication modes play in aesthetics, perception, consciousness, and the sense of self. This is a book that will delight scholars and students alike."

Lance Strate, President, Media Ecology Association and Associate Professor of Communication and Media Studies, Fordham University

"This is a path-breaking book—powerful and innovative. It draws a line around Merleau-Ponty, Ong, and McLuhan, and boldly anticipates the next frontier for communication scholars. Digital technologies and electronic media redefine the social fabric as well as the concepts of self, person, consciousness, and the post-global universe. C. Kaha Waite gives us the language and the theoretical tools to imagine ourselves into this new future. We owe her a great debt."

Norman K. Denzin, University of Illinois at Urbana-Champaign

"This is one of those very serious books that is a delight to read. C. Kaha Waite focuses her analytical mind on the screen of individual 'self.' I appreciated the manner in which Waite relieves human communication theory of the Popperian burden of falsifiability by privileging the mythic in discussing the self, media, the screen, and society. The writing is careful, deliberate, thoughtful, inventive, and provocative. Waite is zealous in her phenomeno-logically driven analysis of the communication matrix. While explicating her understanding of media ecology, the author provides a very close consideration of the noetic consequences of print/writing. I recommend this book to all students of human communication."

Frank E. X. Dance, John Evans Professor of Human Communication Studies, University of Denver

Mediation
AND THE
Communication Matrix

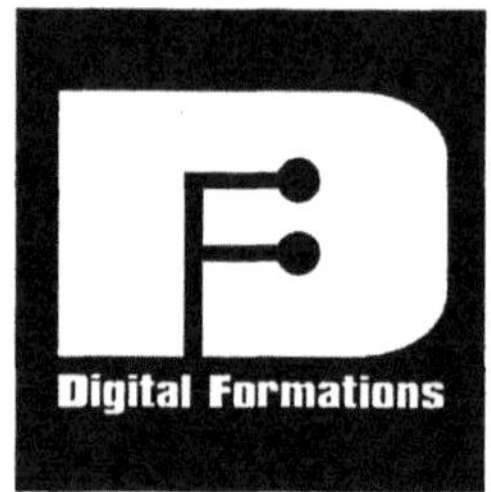

Steve Jones
General Editor

Vol. 10

PETER LANG
New York • Washington, D.C./Baltimore • Bern
Frankfurt am Main • Berlin • Brussels • Vienna • Oxford

C. Kaha Waite

Mediation
AND THE
Communication Matrix

PETER LANG
New York • Washington, D.C./Baltimore • Bern
Frankfurt am Main • Berlin • Brussels • Vienna • Oxford

Library of Congress Cataloging-in-Publication Data

Waite, C. Kaha.
Mediation and the communication matrix / C. Kaha Waite.
p. cm. — [Digital formations; vol. 10]
Includes bibliographical references and index.
1. Digital media—Social aspects. 2. Telecommunication—Social
aspects. 3. Social perception. I. Title. II. Digital formations; v.10
HM1206.W35   302.23'1—dc22   2003016264
ISBN 0-8204-6177-6
ISSN 1526-3169

Bibliographic information published by **Die Deutsche Bibliothek**.
Die Deutsche Bibliothek lists this publication in the "Deutsche
Nationalbibliografie"; detailed bibliographic data is available
on the Internet at http://dnb.ddb.de/.

Author photo by J. Fanelli
Cover design by Sophie Boorsch Appel

© 2003 Peter Lang Publishing, Inc., New York
275 Seventh Avenue, 28th Floor, New York, NY 10001
www.peterlangusa.com

In memory of my father

Paul G. Waite

Contents

Acknowledgments

Though I may once have thought of writing as a solitary activity, I now recognize that this process of putting words to paper is more akin to a shared conversation. My views on these topics have been played out against a rich backdrop of diverse voices. I will always be deeply indebted to the community of scholars who have assisted me in clarifying my perspective on the screen and the social milieu.

There were conversations with Father Walter Ong at Saint Louis University, ongoing discussions with James W. Carey, and e-mail exchanges with Frank E. X. Dance. In addition, I have benefited from my affiliation with the Media Ecology Association. Joshua Meyrowitz has offered his papers and his thoughts. N. Katherine Hayles's book, *How We Became Posthuman*, offers an intriguing alternative to my perspective on the self in the social world. I would like to acknowledge a special debt of gratitude to Norman K. Denzin, University of Illinois. His insightful comments on my scholarship over the years have been deeply appreciated.

In addition, I would like to thank three colleagues who offered detailed comments on the manuscript. Professor Steve Jones of the University of Illinois, Chicago, editor of the Peter Lang Digital Formations Series, who supported the work from the original prospectus to the last rewrite. Professor Joli Jensen of the University of Tulsa, who provided thoughtful commentary on the organization of my ideas. Professor Alfred Kelly of Hamilton College, who has a well-deserved reputation for his systematic critique of written work, reviewed various chapters in depth.

The supportive environment at Hamilton has been very conducive to the best scholarly practices. I feel fortunate to have had excellent assistance from a dedicated library staff, relevant conversations with engaging colleagues, immediate access to technological support, as well as a steadfast commitment from the administration regarding the value of such scholarly projects.

It is my pleasure to thank students Erin Kress for her dedicated assistance with the manuscript and Tara Butler for her interest in and discussions on this topic. Erin's careful attention to details as well as her editorial and research assistance vastly improved the quality of the manuscript. Working with Tara on her senior project gave me the opportunity to teach this topic as I was writing the manuscript. Lisa Rogers graciously provided technical support in formatting the manuscript.

I owe a debt of gratitude as well to those journals that have granted permission to reprint selections from five articles published between the years 1993 and 2002. *Mediation and the Communication Matrix* develops the themes of those earlier publications. Because selected passages from the five articles are scattered throughout the book, I have chosen to mention these articles here rather than include multiple citations in the text.

Critical Studies in Mass Communication, published by the National Communication Association, has given permission to reprint selections from "Toward a Syntax of Motion" (10, 1993, 339–348). *Proteus*, published by Shippensburg University has given permission to reprint selections from "Of Boundaries and Variable-Flex Space" (11, 1994, 18–20). *Cultural Studies* has given permission to reprint selections from "A Syntax of Motion: Time, Space and Television" (2, 1997, 59–70). *Studies in Symbolic Interaction* has given permission to reprint selections of "From Radio to Television: Space, Sound, and Motion" (23, 2000, 113–124). *Studies in Symbolic Interaction* has also given permission to reprint selections from "The Role of the Self in the Social World" (25, 2002, 215–231).

The thoughts that I think, the methods that I employ, and the questions I ask would all be quite beyond my grasp were it not for the inspiring dedication of scholars who came before me. Having extended my thanks to those who directly and indirectly support my work, let me add one final point: the errors and omissions that will inevitably be discovered in this book are my own.

Preface

This book addresses the essential features of the screen and the consequences for the social domain. Though the thesis concerning a new concept of the person has implications for both East and West, the argument presented here does not extend to include a global perspective. The screen offers a unique challenge to American culture. To the extent that the screen inadvertently undermines the very individualism on which American democracy is founded, the dilemma of the screen becomes an American dilemma.

The American colonies envisioned a new form of government that would protect the rights of the individual. That experiment was possible, in part, because as a new country America could escape the traditions of the past in a way that was not possible on the European continent. Today, as American culture confronts unprecedented social changes, there is no opportunity to escape the traditions of our past.

I argue that the concept of American individualism no longer adequately captures our collective experience of the self in the social world. Our laws, as well as many of our shared practices, are based on a concept of the individual that is mysteriously at odds with the current highly mediated environment that envelopes us. Unlike the colonies, we cannot begin the experiment anew.

Many scholars have argued that the emergence of the book contributed to a growing awareness of the autonomy of the person. If the book could contribute to an emerging view of the person as an individual, it is feasible that the screen encourages a very different view of the person. A detailed study of the transition from orality to literacy serves as a guide for investigating the social consequences of the screen. Whereas print privileged the eye over the ear, the screen links eye and ear in kinesthetic motion. The power of the screen resides in its ability to change our phenomenological awareness of the world and, in doing so, to alter our awareness of self and other. The screen creates an altered duration; we experience that altered duration but do not understand it.

In order to discuss that altered duration it is necessary to begin with the phenomenological method, as detailed by the French phenomenologist Maurice Merleau-Ponty. Walter Ong explicitly refers to his own work as phenomenological. Though Marshall McLuhan does not use this term, his work can also be considered phenomenological, given his detailed observations on the subtle ways in which various media alter the human sensorium.

There are many different ways to study the media. Historical, cultural, and interpretive accounts offer insightful theoretical perspectives without ever mentioning the phenomenological method. Yet if I am to successfully explain how the screen alters the social domain, I must begin with perception, with that level of awareness that exists prior to narratives, genres, and stereotypes. The phenomenological method interrogates one's presence in a world; the screen mediates one's presence, altering one's relationship to self and other.

The use of the phenomenological method also challenges one to think carefully about descriptive versus theoretical approaches. Phenomenology is a descriptive method. Joshua Meyrowitz's use of the term *medium theory* emphasizes a way of crafting theories about different media. I remember well a comment made by Father Ong regarding this distinction. He emphatically stated that his work is descriptive rather than theoretical. It is my assumption that he was emphasizing the ways in which description can reveal the world, rather than ordaining the world in advance, which is the risk one runs when theorizing.

It may be that Father Ong's perspective is the right one, that theory runs the risk of influencing one's perspective. Description, in turn, may inevitably be imbued with theoretical implications. My concerns with description versus theory surface after the act of writing. During the process of writing I admit that I am driven by a certain vision, enticed by thoughts that can be given form and shape only through the mysterious act of putting words to paper. I hope that I have succeeded in offering rich description to support theoretical claims and enough theory to illuminate descriptive passages. I will leave that for others to judge.

Mediation and the Communication Matrix turns on four claims, introduced in Chapter One and addressed in greater detail in Chapter Four. These claims are based on a phenomenological approach that investigates the connection between individual perceptual processes and the social domain. In brief, the claims state that one's knowledge of the world is grounded in perception; that one's perception contributes in significant ways to an understanding of the social world; that communication technologies are altering our sense of sight, touch, and movement; and finally, that altering the human sensorium will have consequences for our shared understanding of the social world.

These claims presume that there is a real world that can be experienced and contemplated, that the presentations of the screen are not equivalent to the lived experience of that world. If the presentations of the screen revealed a world no more or less real than any other version—for example, if our experience of standing on a hill watching the sunset were roughly identical to the screen's mediation of that event— the four claims would be moot. Phenomenology presumes that there is a world that can be known through the investigation of perceptual forms. Mediation changes how we perceive the world, and everything follows from that point.

Chapters One and Two identify the challenges of the screen, offer a method, and compare the book and the screen. Chapters Three and Four investigate the essential features of the screen in order to identify the ways in which the screen is changing the social domain. Chapters Five and Six turn to consider the implications for the self in the social world. The necessary comparison is between an American view of the self, as evidenced in the traditions of an earlier time, and an emerging view of the self, evident at the beginning of the twenty-first century. I cannot be certain that I've got it right. I am reassured to think that I am not the first person to be less than absolutely certain of my ideas.

I am also reassured by the fact that many of the ideas in this book are not new. I am working within a tradition; I am asking the sorts of questions others before me have asked. Their work lays out a path for me to follow. My task is to extend that path, if only by a little. Each time I discover yet another author who is working in my domain, I remember that I have it backward: I am working in his domain. There is a rich body of literature that investigates the ways in which the self is situated in the world. I am intrigued by the myriad connections between varied authors, by the sense that I join a large group of esteemed scholars who have thought deeply about these issues. Their work supports my investigation.

There are, I suspect, many different and idiosyncratic ways to read a book. For example, with some books I do not believe it wise to read sequentially from first to last chapter, though I have no idea how I figure out what my approach should be in a particular case. You are invited to read this book in any way that you wish, though I suspect that if you begin with the last chapter you will end with only the rough outline of an idea. I am fascinated by the differences between paragraphs and poetry. There is poetry of a sort in this book, insofar as truth is introduced as myth, argument is supported by description, and description turns on metaphor. Suffice it to say that I have worked to balance the poetic elements with a linear progression of chapters that build sequentially to the end.

Yet the last chapter is more of a beginning than an ending. Chapter Six emerges in response to the previous five chapters, offering an answer as well as a quandary. To the extent that the self takes on a new form, we have yet to recognize the implications of that form. This is the challenge of the twenty-first century for America. Will the American culture ever be able to fully embrace a concept of the self that captures something other than an individualistic perspective? How might we adapt those laws and traditions that are based on an earlier view of the self? There is much more to understand, new paths to be discovered, mysteries to be unraveled. I have entered an ongoing conversation at a particular point. Whether I add something to that conversation or not, I am nevertheless enticed by the possibilities, eager to understand where we are headed, curious to see what emerges round the next turn.

Chapter One:
Method and Matrix

As a result of electronic media, our experience of the world now incorporates space, sound, and motion in unprecedented ways. This powerful and relatively new experience creates an awareness of the world that is vastly different from the experiences of either speech or print. Though the printed text may incorporate leaps in time and space, those leaps are represented linguistically, constrained by the flat surface of the page. Film, video, and computer graphics do what print cannot do: create a sense of movement quite different from ordinary perception.

That altered duration is used not only to tell stories but also to report the news, interview presidents, and explain the latest developments in space technology or baby food. This technological portrayal of dreams and documentaries has altered our shared understanding of the self in the social world. The screen, in its myriad forms, has contributed to an emerging view of the self in American culture that is unique to our time. In order to grasp how this communication technology might alter the role of the self, we must begin by detailing the unique features of the screen. The term *screen* refers to the various iterations of visual media, as exemplified by film, television, digital display, virtual reality, and IMAX. Screen is used as a general term, just as the term *book* is used to indicate many different sorts of publications. Though the screen uses images, it also incorporates sound and motion.

One's experience of speed and movement is vastly different today than in the past. Cars, trains, planes, and escalators as well as video games, film, television, and virtual reality, attest to the fact that technology has created a sense of movement very different from natural physiological rhythms. As Gillo Dorfles explains, "attendance at a cinema or television performance elicits various extremely bizarre phenomena which must be considered adjuncts of our epoch. The spectator is accustomed to register kinetic sequences, developed in time, but in a time—or better, in a duration—which is not that of normal life" (1965, p. 47). That duration is unlike anything experienced in ordinary life, and yet feels real. The viewer, participating in the experience, fails to notice that such "chronological and spatial coercions" (p. 47) would not be possible without the benefit of technology. It is the way in which time, space, and sound are manipulated to create an experience that is both visual and auditory, a sense of motion real and yet unreal. That altered sense of motion is typical of televised ball games and chase sequences as well as commercials, documentaries, and political debates.

The camera reveals to the viewer what print is incapable of spelling out on the flat page. That juxtaposition of sound and space, the altered duration, contributes to the verisimilitude of the screen. No longer a society dominated by print, we have yet to grasp the profound consequences of living in a flux of multimedia. The transition from orality to print restructured awareness of self and other as the work of Walter Ong, Eric Havelock, Jack Goody, and others indicates. Might current communication technologies once again alter awareness of self and other? What Pandora's box is opened with the advent of an electronic mix that privileges not print but image, movement, and sound?

How is the current communication environment altering the self in the social world? Phenomenology can address this question in a way that recognizes the complex interplay of perceptual, technological, and social forms. The phenomenological method will be explained and then related to the medium theory approach. Key concepts introduced in this chapter, such as the communication matrix, will be discussed in greater depth in subsequent chapters. In addition to arguing for the phenomenological method, this chapter explains why some avenues will not suffice to answer our question.

The Phenomenological Method

It is generally understood that video, satellite communications, and the Internet have transformed our awareness of space and time. This point is now largely taken for granted, but the consequences have not yet been fully understood. The corresponding shift in perception is contributing to new forms of knowledge, which, in turn, alter one's relationship to the larger world.

A phenomenological analysis of new media will capture that perceptual shift. As the French phenomenologist Maurice Merleau-Ponty explains, phenomenology is a philosophical approach that "tries to give a direct description of our experience as it is, without taking account of its psychological origin and the causal explanations which the scientist, the historian or the sociologist may be able to provide" (1962, p. vii). There are those who argue that direct perception of phenomena is quite impossible, given the constraints of language and the questionable apprehension of the "real." Merleau-Ponty also understood that the return to perception was a challenging venture. Yet he persisted because of his firm conviction that the western philosophical tradition had misconstrued the role of perception and meaning, had abridged the foundational powers of the body, its ways of being and knowing the world. According to Richard Calverton McCleary, the translator of *Signs* (1964), "carnal self-awareness is the Archimedian [*sic*] point of Merleau-Ponty's philosophy; for the body's presence to itself as both perceiving and perceived provides us with our fundamental knowledge of the consciousness in terms of which the self, the world, and other men are constituted" (p. xvii). If we wish to understand the screen, we must begin by understanding how the duration of the screen alters perceptual processes.

Though I do not doubt the powers of language and social custom to constrain and shape one's perception of the world, I cannot in good faith concede that language and custom are entirely responsible for one's shared experience of self, world, and other. There is something else, something more primary, which escapes linguistic formulas and usurps the constraints of custom. In a fundamental way, "the world is not what I think, but what I live through" (1962, pp. xvi–xvii). That bodily presence grounds one's being in the world in ways that language will never fully capture.

The socially constructed ways in which our culture speaks about self and other, identity and community, do not at first glance appear to be related to how one sees, hears, and feels the movement of a visceral, carnal world. Nevertheless, the perceptual domain is intrinsically related to the social domain. The relationship between the social world and perception is not unidirectional; the influence between these two domains moves both ways. Just as social custom shapes how one sees the world, there are also fundamental perceptual processes that shape social customs. I borrow Merleau-Ponty's method in order to detail the connections between perceptual processes and the social world. I know my relationship to self and other through a primacy of perception that grounds and guides awareness.

There are other phenomenologists who could be enlisted in this project. Edmund Husserl and Martin Heidegger come readily to mind, but neither has investigated perceptual processes as Merleau-Ponty has. His collected writings seek to uncover the origins of the perceptual world. Neither the need to address the dialectic nor the need to bracket experience in order to reveal logical forms constrains him. Merleau-Ponty's approach, more than others, is appropriate for this project because he grasps the primacy of perception. This project depends on understanding the ways in which communication technologies influence perceptual processes.

What is the connection between the human sensorium, the technological interface, and the social world? Taken one at a time, the separate pieces are understandable. When these separate pieces are situated as part of a larger theory, the connections between how one perceives, the communication technologies one uses, and how one thinks about the self in the social world must be detailed. In order to grasp the ways in which the screen is altering the self in the social world, it is necessary to trace the reciprocal influences of perceptual and social forms.

It is understood that social constructions shape how one perceives the world. Peter L. Berger's and Thomas Luckmann's book, *The Social*

Construction of Reality (1967), develops this point in detail. No one doubts that media content has a direct impact on the conceptualization of self and other. Stereotypes, target audiences, concerns with identification and alienation, all speak to the media's power to convey social norms through content. It is easy to note which groups are privileged and which groups are marginalized by newscasts, situation comedies, and dramas. The different roles one plays in community life, as well as the habits and customs of the social milieu, serve to shape perceptions of self and other. Media content displays many of these relationships, whether one is focusing on the family, political figures, gender, or teenagers, to name but a few of the many available subjects.

Our stories mirror the lived world just as the lived world mirrors our stories. It is difficult to discern the starting point, to identify the origins of such shared assumptions about the social milieu. One is left with the postmodern claim that all texts are equally real, or unreal, as the case may be. If one accepts this claim, it becomes increasingly difficult to escape an endless cycle of representations. From a postmodern perspective, we are the story that explains the story that is but a story of the story. On a good day this feels a bit confounding. On a bad day it feels like narcissism.

I want to understand how communication technologies are altering the social domain in America at the beginning of the twenty-first century. I believe there is a real world and that how we think about that world has consequences. I cannot believe that watching a scud missile hit its target on C-Span is equivalent to the soldier's experience in the field as he hears and feels the impact of the explosion. Though the social milieu contributes in significant ways to our perception of the world, that is not the whole story.

It would be naïve to argue that the social domain shapes perception without acknowledging the reciprocal point that perception, in turn, influences the social domain. Merleau-Ponty stresses the profound impact of perception on the social world. According to Elizabeth Grosz, Merleau-Ponty's collective works argue that lived experience "is not only the starting point of analysis but also a kind of measure against which the vagaries of theory can be assessed" (1994,

p. 95). What is needed is a way of understanding how perception influences the social domain.

There are four distinct claims that can guide one in this process. The first claim states that one's knowledge of the world is grounded in perception. The second claim states that one's perception—how one sees, feels, hears, moves—can shape, or structure one's awareness of the social world. The third claim states that communication technologies are altering our sense of sight, touch, hearing and movement. The fourth claim requires a synthesis of the previous three: altering the human sensorium will have consequences for our shared understanding of the social world. Each of these claims will be developed fully in Chapter Four.

The philosophical point that perception provides such a foundation typically carries little weight for most media scholars considering the social and economic impact of a rapidly growing communication industry. There are the challengers and the champions, envisioning both negative and positive scenarios; each is correct in different ways, and neither group need mention the human sensorium or phenomenological foundation to make its point. It is possible to focus on media and never notice the way in which a medium may alter one's perception of the world. Many capable scholars have done so, and their work provides fascinating accounts of economic, sociological, and historical changes that follow the introduction of the printing press, the telegraph, and the telephone, to mention but a few. Yet I am convinced that, in order to more fully understand the social consequences of new media, it is necessary to return to the primacy of perception: to grasp the ways in which communication technologies are altering both our perception of and relationship to the larger world.

Phenomenology is not concerned with the agendas of empiricism. The purpose is not to reduce perception to the basic unit, to dissect or offer causal explanations. From the phenomenological perspective, empiricism misses the profound way in which object and subject are interrelated. One can study the physiology of vision, dissect the neurological features of touch, or track the frequency of auditory patterns, but none of these approaches addresses the issues raised in Merleau-Ponty's later work. In *The Visible and the Invisible* (1968), Merleau-Ponty explains:

> We say therefore that our body is a being of two leaves, from one side a thing
> among things and otherwise what sees them and touches them; we say, be-
> cause it is evident, that it unites these two properties within itself, and its
> double belongingness to the order of the "object" and to the order of the
> "subject" reveals to us quite unexpected relations between the two orders. (p.
> 137)

It is the "unexpected relations between the two orders" that sustains Merleau-Ponty's interest in the gestalt. Our very existence is marked by the motif of figure and ground. As Don Ihde explains in *Consequences of Phenomenology* (1986), "the perceptual model with which phenomenology begins, it is now well known, is a gestalt rather than an empiricist one. By gestalt I mean that the basic 'simple' for phenomenology is a whole/part relation, or better put, a figure/ground relation" (p. 145). Phenomenology argues that, when perception is reduced to a distinct object and separate subject, the heart of the issue is lost: "Here the minimum unit is not the abstract quality, or set of them, which could be built up into entities, but is the interplay of multiple figure/grounds with all the variations which become possible upon that interaction" (p. 161). The empirical and phenomenological agendas differ profoundly in that empiricism does not question the subject-object relationship. As we shall see, this concept of a perceptual gestalt has striking consequences for media studies.

Phenomenology interrogates the subject-object relationship. That is the reason for the emphasis on the gestalt, on the perceptual awareness of a figure-ground relationship out of which meaning arises. Empiricists may choose to analyze perception in terms of distinct qualities. The phenomenologist argues that, in order to understand one's relationship to the world, one must begin with a perception that is both sensible and sentient; hence Merleau-Ponty's claim that the body is both a thing among things, and that which thinks about things. Merleau-Ponty argues that "we have to reject the age-old assumptions that put the body in the world and the seer in the body, or, conversely, the world and the body in the seer as in a box" (1968, p. 138). He rejects both empiricist and rationalist claims.

It is a straightforward point that communication technologies have altered our sense of time and space, our location within a temporal and spatial horizon, in unprecedented ways. This is a phenomenological rather than an empirical claim. The geography of the earth has not changed. Rather, our perception of that geography has been altered; our awareness of the lived world has changed. That change begins first and foremost with a person's awareness that one can now experience the world in new ways. The concepts of near and far, how one thinks about distance, time, and space, characterize a very different experience of the world. This text will detail that difference and explain its significance for the social milieu.

Focusing on the physiology of the eye or the cognitive processes of vision will not serve our purposes. With empiricism, the essential relationship of figure and ground has been rendered unproblematic: vision is located not in the world but in the physiology of the brain. As a way of interrogating the perceptual field, phenomenology seeks to recover the primary features that guide one's being in the world. This venture begins with bodily awareness. As Ihde explains, "the body plays an essential role in perception and the body is a concrete location from which perspectives are had" (1986, p. 60). This venture does not end with the body, because phenomenology recognizes that the body is both a thing and that which thinks of things, both subject and object, implicated in both figure and ground.

The human experience of sight, sound, and touch is profoundly different from the technological extensions of sight, sound, and touch, as evidenced by the telephone, video, or digital imaging. Phenomenology is concerned with perception, with the human sensorium. This philosophical method can be applied to further understand the ways in which communication technology alters that sensorium and to perhaps help one understand the resulting consequences for the social domain.

To sit beside a friend and listen to her voice, noticing her glance and gestures, is nothing like the experience of talking with that friend on the telephone, which by comparison offers far less information. To be present in a physical location, to sense the patterns of sunlight, to perceive the expanse of horizon and feel the wind against one's cheek is nothing like viewing that scene on video. These examples drawn from telephone and video replicate ordinary experiences. There are also forms of mediation that bear slight resemblance to one's ordinary experience: video games, virtual reality, digital imaging, and electronic mail, to name a few. This phenomenological approach, coupled with an emphasis on the medium rather than content, can reveal the perceptual shifts that arise as a consequence of such technologies.

Each of us is born into a physical milieu of speech, sight, and sound, which provides a foundation that grounds our understanding of the world. Media extend that foundation and in so doing alter the human sensorium. Such changes have consequences. Marshall McLuhan argues that media extend the human nervous system. Writing in the sixties, he was not referring to computerized biochips. Rather, he understood that communication technologies serve as an interface that alters one's relationship to the world, not through specific content—though content certainly has an impact—but through the sensorium. Communication technologies alter how the community sees, hears, feels, moves, and speaks. He understood, as few did then or now, that the most important consequences of the interface have little to do with the message, whether that message be story, newscast, video game, or electronic text.

The phenomenologist is not the only one who is interested in perception; the creative artist is also aware of the power of perceptual experience to transform one's awareness of the world. Both McLuhan and Raymond Williams understood this point. McLuhan argues that it is the artist who first discerns changes in the human sensorium, the artist who calls attention to subliminal yet significant perceptual changes in the social environment. Cubism and impressionism are easy to discuss as examples, insofar as these styles are now well understood. Cubism and impressionism captured new ways of perceiving space and light. Referring to the ways in which the screen portrays movement, Williams notes that it is the artist who is most aware of, and interested in, the visual portrayal of movement on television (1974).

Phenomenologist and artist are interested in exploring the ways in which perception contributes to the collective process of sense making. Both also interrogate experience in order to learn more about perception. This approach is easy to talk about when discussing color theory or photography, because the practice of painting or photography requires a vocabulary of terms. The visual artist uses lenses, color, and form for that purpose. It is much more difficult to identify the craft with reference to phenomenology. The tools of the trade for the philosopher are not the artists' lenses, colors, and forms.

How, then, is the communication scholar to use the phenomenological method? It is not enough to recognize that the works of earlier scholars such as Ong and McLuhan incorporate a phenomenological approach. It is possible to appreciate their work and yet not know how to continue that tradition. What tools of the trade might guide one today? The study of technology offers certain conundrums. Investigating perception is not the same as investigating the technological mediation of perception. It is important to recognize the starting point for one's questions, to identify the framework. The medium theory approach offers such a framework.

The Medium Theory Approach

When seeking to understand the complex interplay of social, perceptual, and technological forms, the starting point can be discerned in the half-articulate awareness of the gestalt, of the ways in which figure and ground reconfigure our awareness of the world. The medium theory approach reveals such gestalts by focusing on the essential characteristics of different media. More generally, media ecology investigates the ways in which technologies systematically alter the environment. These approaches sometimes share Merleau-Ponty's interest in perception.

Joshua Meyrowitz provides a detailed discussion of the medium theory approach in *No Sense of Place* (1985), arguing that "the different relationship between content and medium in different media has generally been overlooked by those who embrace the traditional concern with media content" (p. 83). When newspapers, television, or the Internet are viewed as neutral delivery systems, it is possible to ignore the way each of these media influences content. Meyrowitz coined the term "medium theory" to indicate an approach that explicitly considers "the particular characteristics of each individual medium" (p. 16). His point is "that media are not simply channels for conveying information between two or more environments, but rather environments in and of themselves" (p. 16). Meyrowitz's contribution to this tradition was to link medium theory with the work of situationist Irving

Goffman, who studied face-to-face interaction. The approach of the medium theorist is "firmly rooted in systems theory and ecology: When a new factor is added to an old environment, we do not get the old environment plus the new factor, we get a new environment" (Meyrowitz, 1985, p. 19). This approach is also evident in the work of McLuhan and Ong.

The focus on content offers a level of analysis that results in a very different approach. The study of political campaigns, of town hall meetings, of national debates, and other media coverage of candidates provides a wealth of details and offers fascinating commentary on changes in the American political process. Nevertheless, this approach will not provide an insightful analysis of the changing role of the self in the social world because such changes are evidenced in political content, not explained by that content.

A thorough understanding of communication technologies requires that one understand the essential characteristics of each medium. McLuhan argues that the content of media, whether print or broadcast, distracts the viewer or reader from understanding how the specific medium alters one's experience of the world. If one is simply interested in watching the evening news, McLuhan's point will appear irrelevant, if not obtuse. On the other hand, if one is interested in gaining a better understanding of how the media are creating new forms of knowledge and altering the social contract, McLuhan offers a map for how to proceed, though much of the map appears to be written in code. Nevertheless, there are others who clarify and support this venture by investigating the broader historical and cross-cultural context of communication forms.

To examine how media are altering the political process, it would seem most appropriate to investigate the changing nature of presidential debates, the influence of political advertising, the emergence of electronic town hall meetings, and the portrayal of candidates' personal lives in the press. Causal links might be drawn between voter apathy and shifts in the media portrayal of candidates over time. All four items mentioned above refer to specific media content. From political ads to the portrayal of candidates' personal lives, what is scrutinized is the message.

The medium theorist chooses another approach, focusing on the ways in which different media shape public discourse. Less attention is paid to what is being said, and more attention is paid to how the message is contextualized within a particular medium. This approach builds on the premise that the medium shapes the message in important ways. Hence McLuhan's cryptic slogan, "The medium is the message." McLuhan is pointing out the implicit ways in which different media influence content. He refers to this quality of the medium as the invisible environment.

In order to better understand McLuhan's point, consider the different mediums of poetry and dance. It is easy to understand that poetry is nothing like dance, that what is communicated through dance cannot be directly stated in verse. Anyone schooled in the conventions of these forms knows that the power of poetry and dance evolves from the distinctive features of each art. Though both are forms of communication, and both utilize meter and sound, poetry and dance are like different languages.

Imagine how strange it might seem if someone complained that the dance performance was flawed because it did not utilize the poet's words. Or consider the absurdity of arguing that the poem should embody the dancer's form. Such claims run counter to the shared view that these arts are distinct forms that utilize different languages. Dance and poetry explore different approaches to the world: one is kinesthetic, the other linguistic.

Yet, when one considers other forms of communication, the boundaries are not quite so clear. Think for a moment, not about the fine arts but about broadcast and print media. Television and newspapers are, in fact, as different as dance and poetry. These media entail different forms, utilize different structures and conventions, and create incommensurable conversations. In order to understand how these forms of communication differ, one must adopt the perspective of the medium theorist. Less attention could be paid to what is written in the newspaper or spoken on television, and more attention paid to the way each medium shapes the message.

To return to the example of the media and the political process, the medium theorist seeks to understand the essential features of different media. What does television offer that print does not? It could be argued that it is easier for a politician to deceive an audience through print than through television, because the print audience never hears the candidate's voice, never sees the candidate's eyes, is never distracted by a stutter or a vocal pause at an inopportune time, never observes the candidate caught off guard by a revealing question from the press. Print does not provide an opportunity to compare the tone of voice, the message, and the glance. From the perspective of television it appears that print leaves a lot out, but television omits crucial elements as well.

These statements regarding print and television apply regardless of specific content. To extend the example a bit further, a careful analysis of the medium reveals that television rarely offers the detailed perspective of a singular voice. The juxtaposition of time, space and sound, the creation of a duration different from the ordinary, these are the hallmarks of televised discourse. That altered sense of duration creates advantages and disadvantages unique to the electronic medium. The experiences of time, space, and sound are intrinsic to that process.

Havelock, Ong, and Goody focus on the medium when they investigate the transition from orality to literacy. Ong addresses the cultural consequences of print when he details the alienation that follows the transition from orality to script. As Havelock, Goody, and others explain, the form of communication shapes or structures not only the message but the broader environment as well. This insight is evident in Plato's argument against poetry and writing in ancient Greece. Plato sought to break the spell of the muse in order to create a space for critical reflection. McLuhan adamantly focuses on the medium in his exploration of the ways in which media alter the human sensorium.

Though the perspective of the medium theorist is quite different from the perspective of the scholar who focuses specifically on content, it is not necessarily the case that one approach is right and the other is wrong. If the medium theorist's approach were the only way to study communication, then narratives, situation comedies, newscasts, and editorials would be viewed as unimportant. Quite the contrary, the collective stories of a culture are profoundly significant, in that such stories capture cultural identity through social memory and the expression of shared values. Remembering a few of the pivotal moments of this century—the assassination of J. F. Kennedy, Watergate, the *Challenger* explosion—highlights the importance of content. Regardless of the medium, communication technologies have contributed in important ways to defining moments of American history. There are the stories that move us and the stories that offend us, the ways in which the media inform and misinform, perspectives that are included and perspectives that are omitted. Though McLuhan's work indicates that studying content is missing the point, it is not my intention to suggest that content should never be studied.

Both perspectives on the media, the focus on content and the approach of the medium theorist, offer distinct advantages. To suggest that one avenue is correct and the other is misguided is simplistic. These two perspectives, the focus on content versus the focus on the medium, are not entirely separate. Writing in *Rhetoric, Romance, and Technology* (1971), Ong argues that medium and message interact, are interdependent: "In the last analysis, the medium is not even a medium, something in between" (p. 290). Perhaps one day these different ways of looking at the media will be unified into a more comprehensive perspective. There is no denying the power of stories to challenge, illuminate, and transform. The construction of social identity is influenced by portrayals of different groups.

There are also legitimate reasons for studying the medium independently of specific content. It is challenging to detail the implications of what McLuhan referred to as the invisible environment: to identify the ways in which the medium of television alters how all televised stories are communicated or to detail the ways in which print, as a technology, alters social memory. With each new communication technology, the media environment is altered. Different media privilege ways of knowing that are not equivalent to one another.

Add television to a print-dominant culture, and the result is not simply books plus television. The screen provides a kind of experience that was impossible with books, usurping the print-dominant culture in unforeseen ways. The role of books was challenged by television, just as our understanding of television has been altered by computers. As McLuhan and Parker explain in *Through the Vanishing Point* (1968), "each new technology or extension of our physical powers tend to create new environments" (p. 251). Such concerns are best addressed by asking what is unique about the medium.

If the goal is to understand the ways in which communication technologies are altering the role of the self in the social world, then an analysis of specific content will provide, at best, a partial answer. The focus on content provides a wealth of available details. To refer to the Nixon-Kennedy debates, *Leave It to Beaver*, Edward R. Murrow, or the Rodney King incident, is to explore elements of our social history as revealed through a specific medium. It has been argued that the media portrayal of political candidates' personal lives challenges accepted notions of what is off limits, revealing information that would have, in an earlier era, been considered private. This analysis of content is appropriate but does not pursue the question far enough. It is not television that alters collective notions of what information should be revealed in the public forum. Though the media provide the means to reveal such information, a fundamental shift in attitude is also required.

In an earlier era the media could have reported on the personal details of politicians' lives but did not. One might answer that now the public demands such information, but that view does not offer an adequate explanation of the shift in public and private discourse, evidenced in the second half of the twentieth century. Examining television programming will not provide the necessary vantage point. First and foremost, what is needed is an understanding of how the distinctive features of various media are related. Drawing on history, sociology, and philosophy, the medium theory approach explores different media environments.

It is the systems approach, described by Meyrowitz and evidenced in the work of Ong and McLuhan, that provides the necessary framework for exploring the consequences of communication technology. A focus on specific content will not explain how communication technologies have extended the senses, altered the social fabric, created new forms of knowledge, and challenged our understanding of self and other. Let's presume for a moment that one is interested in understanding the impact of television on American culture. This broad and important question can be addressed in a variety of ways. But from the perspective of a medium theorist, what came before television and what comes after television are both important. The book and the computer, comics and video games are significant. Obviously television is but one part of a complex media mix. What is needed is a way of identifying the particular biases of the current environment, which cannot be reduced to just television.

To investigate how the screen has altered American culture is to question the relationship of television to other forms of communication. The intention is neither to ignore nor dismiss content as unimportant. Rather, the intent is to discern the patterns displayed through one form of communication, as opposed to another form—for example, dance and poetry, print and screen, or telephone and face-to-face conversation.

There are many different ways in which scholars focus on the medium. Meyrowitz adopts a sociological approach, investigating how television alters social norms and conventions. Elizabeth L. Eisenstein employs a historical approach, detailing changes in economic and social patterns over time. Havelock and Ong utilize a cross-cultural approach, detailing the emerging cultural changes that arise with new forms of communication. Ong refers to his work as phenomenological. McLuhan's work is also phenomenological, though he never uses the term. Such approaches offer perspectives that transcend discussion of specific content. In addition, these works share one other feature: whether sociological, historical, or cross-cultural, they detail the ways in which communities are changed by the emergence of new communication technologies. All agree that new forms of communication are capable of altering everything from commerce to consciousness.

The decision to focus on the medium is the first step. The second step is deciding specifically what elements or features of the medium will be investigated. In this text, sociological, historical, and cross-cultural perspectives play but a small part. The dominant focus will be on the senses: on perception and the ways in which new media are altering perceptual awareness. As the philosophy of Merleau-Ponty reveals, perception is more than vision, and vision is more than the neat object, outlined like a clear note on a distinct background. Seeing is much more, and much more mysterious than that.

The perceptual shifts of the screen alter the communication matrix. As we shall see, the human sensorium depends on that matrix, or ground, which is not reducible to discussions of eye versus ear. The visual sense is intrinsically related to sound and touch. Perception refers to the multiple ways in which one receives information through the senses. That includes the element of movement, which contributes to body awareness of space or kinesthetics. Ong investigates the eye and the ear in order to capture the changes introduced with the transition from orality to print. McLuhan details the subtle ways in which various technologies challenge sense ratios, arguing that technology extends the human nervous system. Their work provides a model for the investigation of the screen. Their approach must be integrated into a perspective that addresses the consequences of new forms of communication, consequences that transcend the distinctions of orality and literacy.

Ong and McLuhan

This book is motivated, in large part, by the collected work of Ong and McLuhan. Publishing their scholarship largely between 1950 and 1980, Ong and McLuhan advance a detailed account of how communities are changed by communication technologies. Though their work addresses the impact of electronic technology to some extent, significant innovations occurred in the last quarter of the twentieth century. The potential of digital technologies challenges even sophisticated views of how images and words work. Their respective approaches should now be extended to include the electronic media that are rapidly altering the social fabric and with it the concept of individualism that has shaped American society. Ong's use of the term *secondary orality* and McLuhan's concern with the human sensorium deserve more attention, given the highly mediated world that confronts us.

McLuhan and Ong were aware of each other's writings. McLuhan supervised Ong's master's thesis at Saint Louis University. They both believe that the damaging effects of technology can be mitigated, to some extent, through awareness and understanding. Both focus on orality, writing, print, and television when the computer was still in its infancy. Both advocate a broader, systematic approach to the understanding of media. Both are fundamentally interested in the way in which media change human relationships. Ong refers to his work as phenomenological; McLuhan's investigation of the human sensorium indicates a phenomenological approach as well. Both scholars were also working in the tradition of the medium theorist long before that term was coined, though Ong, in his own words, has described his approach as descriptive rather than theoretical (personal conversation, 2000).

Ong is interested in the transition from orality to literacy. He investigates the way in which the spoken word was altered by the introduction of writing and print. It is no longer possible for most people to know what it is like to live in a culture where speech is the dominant form of communication. Nevertheless, Ong believes it is beneficial to understand the loss that comes with the gain of written forms. Ong argues that the emergence of written forms led to changes in community, consciousness, and self. Decades after the publication of his original work, we find ourselves in a strange world, confronted by innovations that far surpass the distinctions of speech and writing. Yet Ong's perspective is still with us, evident in the concerns that technology will alter the social contract, will alter what it means to be in relationship with others, will alter our humanity.

Ong's perspective is with us in another way as well: his focus on orality and literacy provides a useful framework for understanding how technology alters the self in the social world. Whether one is interested in the transition from speech to the phonetic alphabet or the transition from books to electronic hypertext, the role of the self in the social world remains a concern. Ong understands that new forms of communication inevitably alter one's relationship to self and other in unforeseen ways.

Perhaps McLuhan pursued his investigation of the media because he was confounded by the changes that were confronting him. Drawing on the work of Harold Innis, Gregory Kepes, and Edmund Carpenter, among others, he details a singular and idiosyncratic vision of technology's impact on society. His work has mystified and enticed both layman and scholar since its appearance in the 1950s. Investigating the experiential features of media that few people notice, McLuhan details the ways in which communication technologies alter the human sensorium. Interested in the electric light bulb, typography, trains, and roads, he is acutely aware of everything; for McLuhan, everything relates to something else. He explains the significance of top-less dancers in the same way that he links Shakespeare's *Romeo and Juliet* to television. Using language many would find obscure, he explains how different media are hot or cool, why it is significant that, unlike film, television projects light onto the viewer, and how the alphabet alters our awareness of nature.

He is interested in the changes in scale that new technologies create, as evidenced in a shifting awareness of space and time, and in the social and psychological consequences of electricity, television, comic books, ads, and radio, to name only some of the media on which he focuses. He considers many things to be important forms of communication, such as clothing, numbers, and money. Arguing that we experience far more than we understand, he suggests that it is experience, rather than understanding, that guides one's behavior. By relentlessly exploring the perceptual biases of various media, he seeks to enlarge awareness of how one is influenced by these forms. Believing that profound changes follow the introduction of a new technology, he insists that we are numb to the consequences of new media, because we are mesmerized by our technologies.

Whereas Ong investigates how the technology of writing alters consciousness, McLuhan explores the ways in which new forms of communication interface with the human sensorium. One of the most difficult points to understand in McLuhan's work is his belief that media become physical, social, and psychological extensions of the human nervous system. Hence his title, *Understanding Media: The Extensions of Man*. For McLuhan, media extend our senses and in doing so alter our physiology—a crazy idea until one stops to consider that the human sensorium consists of eyes, ears, touch, and kinesthetics. Technology allows one to see across vast distances and to send the sound of a live voice from one country to another. It is commonplace to share the images of daily life from one part of the world with another part of the world, and in doing so, to perhaps be aware of others' joy and pain. How could such technology not have an impact on the ways we perceive the world?

McLuhan argues not so much for a utopian vision of cosmic unity than for a heightened awareness of the social consequences of communication technologies. His message is difficult to understand, in part because he creates his own language to explain key concepts. It is only when his ideas are placed next to another's work that his message becomes more accessible. It is helpful to place the work of McLuhan and Ong side by side. Ong understands, as well as any can, that an oral culture is distinctly different from our contemporary highly mediated culture. Ong's work creates a context for McLuhan's message.

McLuhan's work has been dismissed as deterministic because he argues that technology has a profound impact on how a culture experiences and thinks about the world. Undoubtedly, the criticism of technological determinism could be made about this project as well. I hope to address that criticism in advance or at least contribute to the debate by questioning what is involved in the concept of technological determinism.

Americans do not restrict new technologies; we explore them, like explorers of earlier times who conquered new frontiers, creating maps as they went. Sustained by optimism and enthusiasm, we eagerly investigate possible applications, figuring out how to adapt the technology to our purposes. When television was first invented, it was referred to as a rich man's toy, and many wondered what people might do with such a device. Now it is a fundamental feature of American life, and its influence has exceeded our wildest dreams. The entrepreneurial nature of American culture insures that new inventions are put to use, even if it is unclear at first exactly how they might be used.

Today we have not just television, not just text, not just computers, but hybrids of all of these. The multimedia monitor provides image, sound, and text. Cable and telephone lines provide immediate access to sites all over the world. It is difficult to grasp the implications of these hybrid technologies, though it is clear that what we are dealing with now is not like television, is very different from telephones, and is nothing like books.

Critics debate the merits first of this technology and then of that one, but commentary on individual technologies will not provide a systematic account of the ways in which such innovations are altering the social domain. What is needed is a way of capturing the shifting figure and ground that creates the new gestalt. My primary concern is neither technological application nor human expectation but the subtle and profound ways in which application and expectation each influence the other. In the application of new technologies, human expectations are inadvertently altered. In turn, our expectations of what is possible shape the ways in which new technologies are applied.

The decision to use a specific technology in one way, as opposed to some other way, can have profound consequences. The legal precedent of the *scarcity rationale* had a direct impact on the development of radio in this country. As a consequence, the First Amendment has been applied very differently to print and broadcast media. Though television is now recognized as a commercial medium, according to Harry Matthei, there was a time "when all TV licenses were strictly 'experimental' – i.e., noncommercial" (1999, p. 155). Early violators of the noncommercial rule drew fines and threats of license suspension. The Supreme Court's decision to rule the Communication Decency Act of 1996 unconstitutional had a direct influence on the future use of cyberspace. Each of these human actions shapes communication technology to one agenda or another.

In turn, communication technologies also challenge human expectations, altering the nature of business, politics, education, community life, and childhood. When a community adopts a new technology, the practices of that community will be altered in unpredictable ways. It is not possible to be cognizant of such changes in advance. Who might have guessed the extent to which the telephone would eventually alter when and how people speak to each other? The inventors who believed that television was merely a rich man's toy could not have predicted the power of television to shape public discourse. When the military began development of the Internet as a way to share information, could they have imagined an Internet chat room where people exchange other kinds of secrets? Each innovation brings both gain and loss. My purpose is to understand that gain and loss more thoroughly in order to grasp the ways in which communication technologies alter the role of the self in the social world.

A strict adherence to technological determinism would unnecessarily simplify the complex balance between human expectation and technological application. Human expectation can have a direct effect on what sorts of applications a community chooses to pursue. The regulation of broadcast media by the FCC has resulted in clear distinctions being drawn between print and broadcast. The development of the Internet has been influenced by the fact that it is largely unregulated. In turn, new technologies also have a dramatic effect on one's understanding of what is possible. A print-dominant culture understands the world very differently than a preliterate culture.

With each new form of communication, a niche is created. People respond to that niche in unforeseen ways. Technology influences not only expectations but perception, cognition, and social life. The lack of a direct causal relationship does not negate the power of such influence. Goody responds to the charge of determinism in *The Power of the Written Tradition* (2000) by arguing that "the products of technology are external only in a formal sense, since they are formed by humanity and in turn form humanity" (p. 136). Drawing on the example of writing, he explains that, "with a pen in our hands, we are different than when we carry a sword or work on the loom; we have different roles that structure our perceptions" (p. 136). Just as the use of tools influenced human evolution, tools of a more complex design influence us today.

The question is not whether we are influenced but how we are influenced. Does the influence extend beyond expectation to social roles as Goody suggests? Does that influence extend to how a community thinks about self and other, or is such influence limited to how forms of knowledge are created and utilized? In *Television: Technology and Cultural Form* (1974), Williams argues for an emphasis on human action rather than determinism. His point is that new inventions are created with certain intentions in mind and that such intentions predict the ways in which a particular medium will be used as well as the influences of that medium. His position fails to account for the emergent, sometimes surprising applications of new inventions or the way those surprising applications can usurp the best intentions.

Who might have anticipated that the telephone would have an impact on the activity of writing letters or foreseen the ways in which computer technology undermines long-distance telephone service? How a community chooses to use a new technology is related to the emergent features of that medium. It is feasible to create links with hypertext that would be quite impossible with the traditional book. Television can reveal features of the social world that could never be adequately displayed in newspapers. Determinism simplifies the complex reciprocal influence of technology and society. The question is, how are we influenced by new media? Because it appears that we are influenced in unpredictable ways, it would be helpful to know more about the nature of that influence.

Take the phenomenological method, using the work of Ong and McLuhan as a model, and incorporate the medium theory approach to the study of technological innovation. What do we have? A way of thinking about technology that relates the essential features of different media to the perceptual field. Ong and McLuhan address the essential features of orality and literacy in terms of the ear and the eye. I suggest that new forms of communication transcend the distinction of ear and eye. In *The Gutenberg Galaxy* (1962), McLuhan explains that the senses can be "embedded or outered in material technologies," and when "so outered, each sense and faculty becomes a closed system. Prior to such outering there is entire interplay among experiences" (p. 265). I suggest a different way of thinking about the human sensorium: that technological mediation can alter the ground or matrix out of which sensory experience arises.

The Communication Matrix

The term *communication matrix* identifies not the eye or the ear but that complex human sensorium of speech, vision, hearing, gesture, touch, and kinesthetic processes. The communication matrix includes those subtle and profound ways in which humans interact with the phenomenal world. Any discussion of the senses must acknowledge their interrelationship. That interrelationship can be captured by the term *matrix* to indicate the way in which each of the senses confirms and sustains the others. The noun *matrix*, from the Latin word for "womb," refers to an enclosure within which something originates or develops.

This is a phenomenological rather than a physiological point. I am not suggesting changes in neurological structures, though one might consider making that argument. As Merleau-Ponty explains, one is both a thing among things and that which looks at things. It is the reciprocal nature of experience, what Merleau-Ponty refers to as the intertwining, that reveals the ground or matrix. My sense of the world, my vision of the horizon, is not some separate and distinct event. Because I am both of the world and in the world, each time I engage the world I am caught up in that engagement. Change the way I interact with the world, and I am changed as well. I cannot privilege one sense over another without altering everything in the process. If I close my eyes to listen to a concert, the music becomes tactile, perhaps kinesthetic; I feel it as well as hear it. The senses are deeply interrelated: vision, hearing, touch, movement, and scent confirm my presence in the world as well as the world's presence in me.

One's experience of the world is grounded in a horizon that is immediate, personal, and knowable. Change the nature of that experience in subtle and profound ways, and relationships will inevitably be affected. Though it is not possible at this point to alter the fundamental character of the human ear, eye, or voice, it is possible to extend the reach of one sense over another through technological enhancement. Consider the powers of amplification provided by the microphone or the camera's lens. Mediation alters the ways in which we experience the world and, in doing so, alters the communication matrix.

This understanding of the communication matrix is based largely on the work of Merleau-Ponty. According to Grosz, who details Merleau-Ponty's perspective in *Volatile Bodies* (1994), "in lived experience the senses interact, form a union, and yield access to a singular world" (p. 99). Merleau-Ponty, in *Phenomenology of Perception* (1962), writes that "the sight of sounds or the hearing of colours come about in the same way as the unity of the gaze through the two eyes: in so far as my body is, not a collection of adjacent organs, but a synergic system" (p. 234). He contends that one's physical presence provides the foundation for knowledge of that world; he writes that "my body is the fabric into which all objects are woven, and it is, at least in relation to the perceived world, the general instrument of my 'comprehension'" (p. 235). Perception is foundational for Merleau-Ponty because "perception is not a science of the world, it is not even an act, a deliberate taking up of a position; it is the background from which all acts stand out, and is presupposed by them" (pp. x–xi). The communication matrix does not refer to media, as in "media matrix" (Meyrowitz,

1985, p. 339). Rather, the use of the term indicates the primacy of a perceptual figure-ground, out of which meaning arises.

Merleau-Ponty's writing gives form and shape to what one may already know in some inarticulate way. Vision is a kind of touch; touch is a way of knowing that precedes other forms of knowledge; the gaze interrogates and, in doing so, is in turn interrogated by a world that is first and foremost a blooming cacophony of movement, sound, and vision. This is the way we are born into the world. This is our human inheritance. As writers from William Blake to Ludwig Wittgenstein have eloquently pointed out, other forms of knowledge are not available to us. If a lion could speak, would we be able to understand? Can we know the winged world of a bird in flight? The human sensorium forms a communication matrix in which the senses confirm one another, coordinating a perceptual language that gives rise to an understanding of world, self, and other.

The writing of Ong and McLuhan offers an insightful framework for exploring the communication matrix. These scholars begin by asking how the ear differs from the eye, how the auditory experience of speech differs from the silent experience of reading, how reading differs from watching television. Ong and McLuhan investigate the ways in which perception, technology, and community are related. Speech is the first form of communication, the foundation on which other forms are based. Writing inevitably alters one's experience of speech, contrasting the spoken word with the letters on the smooth page. Images also shape awareness, from the earliest hieroglyphs that evolved into written forms, to the moving images of early films, to the digital images of the screen.

McLuhan understands that the inherent biases of different media alter the human sensorium, that media serve as extensions of our senses. McLuhan writes that "every new technology thus diminishes sense interplay and consciousness, precisely in the new area of novelty where a kind of identification of viewer and object occurs" (1962, p. 272). Technology is not just out there, media are not separate from our ways of hearing, speaking, feeling, and seeing. Whereas McLuhan argues that a new technology diminishes the interplay of senses, I suggest that interplay—the relationship of sight, hearing, touch—can be heightened, certainly altered, but not necessarily diminished.

Beginning with the basic features of sight, hearing, and touch, humans have created complex communication systems to express, portray, and illustrate their experiences. These complex systems alter the ways in which the individual experiences the social world. Even the distinctions of speech, image, and text introduce unavoidable ambiguities. Images can mean photographs, graphic renderings, television images that incorporate sound and movement, or iconic forms. Text can mean newspapers, novels, web-based linked text, or poetry. Despite these ambiguities, the focus on speech, image, and text highlights central features of media that correspond to the human use of sight and hearing.

One's first experiences of the social world are through speech. Speech includes those varied linguistic and paralinguistic practices that situate one in relationship to others. Despite the layering of hybrid media, speech remains foundational. Without speech there is no text, subtext, or communication with others. Ong developed this theme throughout his work, focusing on orality and literacy in order to detail social and psychological shifts in awareness created by the emergence of writing and print. Without an understanding of linguistic forms, there can be neither communication nor the media that facilitate communication. In this way speech can be viewed as the genesis of all technology. One's first and most provocative experience with language is through the spoken word in concert with others. As Frank E. X. Dance argues, "the spoken mode is developmentally central" (1979, p. 203). It is that formative experience that shapes each of us. It is the phenomenological features of that first language that are altered by the use of various media.

Those phenomenological features include the presence or absence of the other, and the broader context of speech that includes gestures, eye contact, breath, and intonation. With face-to-face conversation, one feels the gaze of the other as she speaks. By contrast, one can stare intently at a speaker on television, film, or video and never feel the speaker's gaze: an inconsequential point except for the primary importance of eye contact in face-to-face conversation. This is one of many elements that are altered with mediated communication. Through meaningful dialogue, self and other are acknowledged. Mediated forms of communication build upon and alter that experience, adding layers of complexity. Spoken language, or orality, is also the starting point for the successive communication revolutions that have contributed to the current environment. Writing, printing, and electronic forms have been contrasted and compared to the fundamental qualities of speech in order to better understand changes that technological innovation creates. To some extent, this way of proceeding can draw on the biases of the past to understand the future. As the varied images of the screen have challenged our print-dominant culture, confusions arise concerning words and images. In addition, the call for media literacy has not provided a thorough understanding of the inherent biases the term *literacy* implies nor an awareness of the ways in which new media move beyond literacy.

Such confusions make it more difficult to understand how the book and the screen differ. The screen is not merely images. Ideas are not communicated only through words. Media literacy, as Meyrowitz explains, requires more than a critical understanding of content. As long as one thinks of the screen as a wayward version of a book, it will be difficult to grasp the fuller implications of the screen. The implications of the book and the screen are fundamentally different. In order to understand what is unique about the screen, it might be helpful to first address confusions about media literacy as well as confusions about how words and images work.

Of Books and Screens

Many have argued for the importance of media literacy without understanding that a broader definition of media literacy is needed to clarify how communication technologies are altering the self in the social world. Typically, media literacy advocates encourage the sophisticated viewer to recognize the agendas, stereotypes, and biases of the dominant media. Other types of media literacy are called for that address the medium in addition to specific features of media content. Meyrowitz (1998) argues for three forms of media literacy. The third of these, *medium literacy*, illustrates the relationship between the communication matrix and the social domain.

As Meyrowitz explains, the first is *content literacy*, which is what most people mean when they argue for media literacy. Content literacy focuses on the narrative, newscast, or dialogue, on specific features of what is being communicated. The second is *grammar literacy*, which investigates the forms, grammars, or structures that different media use to communicate the message. For example, zoom and wide-angle shots are, loosely speaking, grammars or organizing features of video, whereas the inverted pyramid of who, what, where, and when, has long been an organizing feature of print journalism. The third is *medium literacy*, which focuses on the ways in which a particular medium contributes to social organization. For example, television altered collective notions of scale, or space, as events in one nation or region were broadcast live all over the world.

Consider the differences between a printed book and computer-based text. The differences these forms of communication exhibit—the fixed words of the printed page, compared to the nonlinear links of hypertext—can influence the literacy of the reader in at least three different ways. Based on Meyrowitz's analysis of the three types of media literacy, the difference between fixed and floating text influences (1) how content is interpreted, (2) how distinct grammars structure the message, and (3) how different media alter perceptual and social processes. These types of media literacy are not entirely separate. Content is structured through the grammar of a particular medium, and grammar, in turn, shapes awareness of the social domain.

In order to discern the ways in which communication technologies alter the social domain, the essential features of the medium must be identified. Ong's emphasis on orality and literacy stands as one example of this approach. This is the mystery of the ear and the eye, the conundrum of spoken and visual languages. What possible connection might exist between the earliest forms of orality and the highly visual culture in which we now live? If spoken forms provide the genesis of mediated communication, the dominance of the eye may more fully capture the current media mix. The emphasis on spoken language cannot address the role of vision. An investigation of visual forms of communication leads in a very different direction.

From the pervasive billboard to the glossy magazine ad, the fluid video image to the stereoscopic experiences of film and virtual reality, we live in a world of mesmerizing images that operate with a logic strikingly different from that of linguistic forms. Part aesthetic, part commercialization, part iconographic culture, these dynamic forms cannot help but impact the human sensorium. The intense visual nature of this iconographic culture, whether those visual forms be neon, digital, or video, is vastly different from the visual experience of the fixed page. We have the ability to integrate text and image, to create diverse fonts and designs, to digitize a vast and compelling array of images. Nevertheless, the focus on a visual culture can mislead those who seek to understand the social significance of the screen.

Because television has been widely available for approximately half a century and constitutes one of the older mediums, this technology can illustrate some of the confusions concerning visual images. Though this medium utilizes speech, images, and text to a small degree, critics often focus on the image. Comparing television and print, critics claim that print is rational, substantive, logical, and capable of sustained discourse. By contrast, many have argued that televised images are impressionistic and emotional, not conducive to sustained engagement. Many points follow from this perspective on words versus images: television viewing reduces children's attention span; achievement scores are negatively affected; family life is compromised; violence is encouraged; the political process is reduced to a war of personalities; voter apathy is encouraged; public discourse is harmed, and cultural forms are sacrificed to the vulgar. Nothing less than education, family life, the political process, and civic discourse is affected.

Critiques of television in the 1980s and 1990s compared the medium to print. The comparison was rarely positive. In *Amusing Ourselves to Death* (1985), Neil Postman writes that "under the governance of the printing press, discourse in America was generally coherent, serious and rational...under the governance of television, it has become shriveled and absurd" (p. 16). Concerns with television often focused on the negative consequences for education: "In 1991, Donald Stewart, president of the National College Board, blamed the SAT score decline on 'too much TV, too many videos, a decreased amount of time spent reading'" (Bianculli, 1992, p. 185). Norman Corwin, concerned that television trivializes the news, writes, "a public fed on prattle may suffer diminished capacity to weigh the news" (1989, p. 28). Kathleen Hall Jamieson, in *Eloquence in an Electronic*

Age (1988), writes that television has the "power...to argue visually in ways that defy traditional evidence" (p. 122). Michael J. O'Neill writes, "Unfortunately, television is an impressionistic medium that marshals images and emotions rather than words and reasons" (1989, p. 10). Such commentary reflects a shared belief that the visual quality of television has the power to create a society less capable of critical thought.

Such critiques do not fully address the primary features of the medium. Television utilizes images in a way that is simply not possible with print. Furthermore, the screen cannot be reduced to mere images. Though its impact on American life has been profound, the fallacy of the single cause suggests that the social, political, and psychological changes evidenced in the latter half of the twentieth century cannot be attributed solely to television, no matter how pervasive and powerful the emergence of that technology might have been. Lamenting that television is not like books does not address the fundamental difference displayed by these two media.

Television is more than images and emotion. Print is not necessarily inherently logical. In his later work, Wittgenstein (1958) argues that previous notions of logic were too tightly constrained, that much of language use cannot be adequately captured by the logician's algorithm. There are many different ways to use language, analogous to different games. People recognize that there are different ways of speaking for different situations. We move freely from one language-game to another without confusion or constraint. Poets invent new language-games for their pleasure without the restrictions of logical forms. It is not that some of our most important ideas disavow logic, but rather that the logic of our utterances is far more diverse than imagined. Logic is not the exclusive domain of print any more than images are exclusive to television. Such false dichotomies obscure our understanding of both mediums. Insofar as images are meaningful, there is, by virtue of that meaningfulness, an implicit logic.

No longer a print-dominant culture, we experience public discourse to a large extent through a mix of images and words that operates very differently than print. As a visual medium, television foreshadowed future possibilities evident with the use of digital images and multimedia. Now computer-based technologies are merging the distinct features of television, print, and telephone as Web-based multimedia incorporate images, text, and sound.

The champions of television advocate new forms of literacy, recognizing that television offers different ways to structure and access information. David Bianculli (1992) argues that, with careful attention to programming, children can and do learn a vast amount from the television. In addition, Bianculli understood a full decade ago that the real advances in classroom television were linked to the possibilities that interactive media provided. Electronic media alter how subjects are taught and how students learn, providing opportunities to connect with others in ways that were previously impossible. The computer provides a new perspective on television.

McLuhan argues that "today technologies and their consequent environments succeed each other so rapidly that one environment makes us aware of the next" (1964, p. ix). The newer technology in this case is the computer. Changes that the computer created provide a vantage point from which to explore the preceding technology. Both television and computers present information in a visual format. In one case, visual images are considered to be detrimental; in the other case, visuals are an intrinsic part of the program. Examples that come easily to mind are computer-aided design, programs that teach calculus, video games, and virtual reality. But this point obscures what is most important: as a result of the screen, our experience of the world now incorporates space, sound, and motion in unprecedented ways.

Rudolf Arnheim challenges the view that the screen portrays mere images with a systematic analysis of visual thinking. Arnheim understands that the screen reveals a world in motion. As he explains in *Art and Visual Perception* (1954), "the motion picture has broadened not only our knowledge but also our experience of life, by enabling us to see motion that [was] otherwise too fast or too slow for our perception" (p. 385). Arnheim understands that the camera uses motion in at least two different ways: "it presents a picture of motion, such as that of a runner, or a dancer" (1957, p. 168), but it also captures the passage of time—

> We know that the sun changes its place in the sky; but when a film, in condensing a day to a minute, shows the play of rapidly moving shadows interpret the plastic relief of architectural shape, we are made to think of light as

> a happening that assumes its place among the other productive motions of
> daily life. (1954, p. 385)

The camera condenses the blooming of a flower to a matter of seconds, captures the face of the actor moving from grief to joy, reveals the juxtaposition of light and dark, the sky in morning and then night, portrays the disembodied voice framed by shadows moving across a sunlit window.

Temporal sequence reveals the evocative character of a world captured through motion and sound. It is this altered duration, coupled with contextual nonverbal cues, that many have dismissed as merely images. As Arnheim explains, "Many educators and psychologists are still reluctant to admit that perceptual thought processes are as exacting and inventive and require as much intelligence as the handling of intellectual concepts" (1965, p. 2). His view on perceptual thought processes is confirmed by the advent of computer imaging that facilitates the exploration of ideas through iconographic forms.

Beginning with Plato, there has been a lengthy intellectual tradition that privileges thought over perception. Thought is identified with linguistic processes, and perception is identified with images. Arnheim challenges this bifurcation, arguing that "visual perception...is not a passive recording of stimulus material but an active concern of the mind" (1971, p. 37). Perception and thinking are not separate activities; one can think in images as well as words. As Arnheim explains, "the kind of process observed in logical thinking occurs at the perceptual level also" (pp. 39–40). The performing and fine arts explore diverse forms of thought through the acoustics of music, the images of the visual arts, and the movement of dance. While some might argue that the screen presents mere images, it is more difficult to argue that the fine arts present mere images, or that such images do not contribute to conceptual thinking.

As a conceptual and intellectual activity, thinking has generally been identified as a linguistic process. Arnheim challenges this view, arguing that "thinking treats space and time, which are containers for being, as the structural categories of coexistence and sequence. Both of these categories can be represented in the spatial medium of visual patterns" (1971, p. 129). As a medium, the screen provides for a kind of intellectual activity that operates very differently from the fixed succession of words on the flat page. To the extent that the screen emerges as a dominant technology, our understanding of conceptual thinking may shift to reflect the influence of that electronic technology. Arnheim's work explores the multidimensionality of thought in order to address the complex ways in which perception and thought are interrelated.

Sentences must exhibit a logical structure in order to be meaningful. Moving images must also display a structure based on a duration which situates and confirms the details. If the images of the screen did not, in some way, confirm our everyday experience of duration, of gesture in time and space, those forms would be meaningless rather than captivating. In each case, it is the context that is crucial. Ong's distinction between written and oral language helps explain the significance of that context:

> Written discourse develops more elaborate and fixed grammar than oral discourse does because to provide meaning it is more dependent simply upon linguistic structure, since it lacks the normal full existential contexts which surround oral discourse and help determine meaning in oral discourse somewhat independently of grammar. (1982, p. 38)

Print establishes context through grammatical structure. Oral discourse establishes context through those elements print omits, such as gesture, eye contact, and nonverbal cues. The screen builds on that "normal full existential context." The context of one's words, whether spoken or written, and the context of images function very differently.

Words are capable of ordering one's thoughts in sequences of before and after, if and then, past and future. By freezing thought on the smooth page, print reinforces this sequential quality of language. Visual images create a gestalt that functions very differently than words. Writing about the perception of visual images, Arnheim explains that "the properties of any part of the visual field must be seen in constant relation to corresponding properties of the field as a whole" (1971, p. 40). Images are experienced with a simultaneity that is distinctively different from linguistic forms. But that simultaneity does not necessarily imply a lack of conceptual thought.

Both images and words contribute to an understanding of the "context" about which Ong writes. Consider poetry, a form that is typically communicated through print. If television is problematic because it uses images and does not communicate information logically, then what of poetry? Good verse makes ready use of images, disavows strict logics, invents new forms of expression, and generally subverts grammatical rules. Print also relies on the image, the evocative sign, the narrative, and the symbolic.

An epistemology that equates words with thought and images with emotion will be incapable of addressing the essential features of the screen. The advent of printing encouraged the divorce of word and image. Early manuscript culture did not distinguish between words and pictures. The illuminated manuscript made pictures out of letters, and letters out of pictures, as ideas were expressed in a variety of forms. The manuscript was not constrained by the traditions of print. The book, by contrast, turned words into things. As Ong explains, words become objects on the page rather than actions to be experienced. Images, pictures, charts, and diagrams serve as mere illustrations of the word. Ong argues that thought is nested in speech, a lovely turn of phrase that emphasizes the primacy of spoken language. With the book, thought becomes nested in the printed word. The bias of print separates not only eye and ear but also word and image. The screen challenges the bifurcation of word and image.

Lived experience does not so easily divorce word and image; by contrast, phenomenology seeks to integrate what is seen and said, questioning the primacy of language and arguing for a perception that grounds thought in bodily awareness. The human condition is grounded in more than words. We are not first and foremost speech. There is the hand that reaches and the eye that searches for confirmation. Thought cannot reside only in words, nor can images serve merely as illustrations. Words and images both support thought.

The unique features of the screen can best be understood through the integration of what is seen and said. Arnheim refers "to the belief—common to philosophers, who are men of words—that concepts can be produced only by means of words" (1965, p. 6). Concepts can be expressed through both visual and linguistic form. Ask the artist, scientist, or poet: creative activity cannot be restricted to just words or just images. The screen provides an intriguing counterpoint to the bias of print, challenging one to rethink the relationship of words and images.

For Merleau-Ponty, images and words share a common origin. Both images and words are grounded in perception. Without a body located in a world, without awareness of space and time, without a perceptual horizon, there would be no way of understanding either words or images. Merleau-Ponty recognized that perceptual forms constitute a syntax; that perception provides the basis for language just as language signifies and refers back to that perceptual world. In *Signs* (1964), he writes:

> We must therefore recognize that what is designated by the terms "glance," "hand," and in general "body" is a system of systems devoted to the inspection of a world and capable of leaping over distances, piercing the perceptual future, and outlining hollows and reliefs, distances and deviations—a meaning—in the inconceivable flatness of being. (p. 67)

Merleau-Ponty interrogates the perceptual realm prior to language, out of which language arises: "If it is characteristic of the human gesture to signify beyond its simple existence in fact, to inaugurate a meaning, it follows that every gesture is *comparable* to every other. They all arise from a single syntax" (p. 68). The assumption that thinking and sensing are separate and distinct categories makes it difficult to understand the shared perceptual domain of words and images.

Inexorably, the screen moves one closer to Merleau-Ponty's perspective. Unlike print, the screen weaves images and words into a seamless portrayal of the world that can be both fascinating and oddly alienating. The screen reveals a gestural language that is visual, auditory, and kinesthetic. Merleau-Ponty argues that perception provides the basis for language; the screen mediates perceptual forms. We see the world close up, in motion, fascinating in its familiarity, and alien in ways that are difficult to fathom. The landscape is reconfigured, the human voice is modulated by surround sound that intensifies the experience of both eye and ear, the experience of space and time is warped by the camera's perspective. The varied perspectives of the screen display the potential ways in which perception can "inaugurate a meaning."

Why then does the screen leave one feeling oddly alienated? Could McLuhan's suggestion that we are numb to the effects of a new technology be correct? Though most might agree that American culture is influenced by technology, the question is, how are we influenced? Historically, America was shaped by a frontier that—without regular mail, radio, or telephone—was far more isolating than anything we know today. What language might adequately capture today's technological frontier that offers new forms of both isolation and connection? What are the consequences of these new forms, which result in a very different way of knowing the world?

Caught up in unprecedented changes, it is difficult to discern the outlines of current transformations, much less discern the shape of a future that is beyond our grasp. Computers provide a new perspective on television, just as television did for film, and film did for books. Older technologies become art forms, to be examined and explored in depth. The past can be studied; the future can be imagined, but it is difficult to discern the present.

McLuhan suggests that the artist is capable of picking up "the message of cultural and technological challenge decades before its transforming impact occurs" (1994, p. 65). McLuhan's faith in the artist reflects his emphasis on the ways in which new technologies alter the sense ratios. The artist is a trained phenomenological observer of perceptual processes. Williams also notes that it is the artist who understands the power of new forms. With reference to television, he notes that the medium creates a sense of visual mobility, contrast of angle, and variation of focus that painters recognize more easily than do communication scholars:

> To most analysts of television, preoccupied by declared or directed content, this is, if seen at all, no more than a by-product of some other experience. Yet I see it as one of the primary processes of the technology itself, and one that may come to have increasing importance. And when, in the past, I have tried to describe and explain this, I have found it significant that the only people who ever agreed with me were painters. (1974, p. 71)

The artist discerns things others miss, is capable of responding to the shifting foreground and background. McLuhan recognized the importance of the artist's perspective in 1964. Williams noted that same point in 1974. Almost four decades later we find ourselves in the midst of an information revolution, confounded by image, movement, and sound. What was true then is all the more true today. Electronic media alter time and space in a way that print never could. What may have been subliminal decades ago is now obvious to anyone who manipulates computer graphics, plays video games, attends movies, watches television, or drives by the neon billboards that light the urban landscape. Viewers are drawn empathetically into that felt sense of motion in a way that was simply impossible with print. The defining elements of a print-dominant culture are being transformed by a very different perceptual experience.

Let us trace out the implications of an earlier revolution, in order to discern patterns in this new revolution. The transition from oral to script to print cultures offers a kind of map that can guide our investigation. The book provides a counterpoint for understanding the screen. The transition from orality to literacy created new ways of perceiving and thinking about the world. The organizing principles of a print culture are profoundly different from the organizing principles of the screen culture. A brief review of the transition from orality to print helps one discern what may be coming with the screen.

Chapter Two:
Earlier Revolutions

This chapter addresses the story of the transition from orality to script and the consequent development of printing. Though the digital era is nothing like that earlier era, if examined carefully, the transition from orality to print will provide a way of understanding the implications of today's electronic revolution. There is a story, shared among a select group of scholars, about the transformations that occurred with the transition from orality to writing. This story spans a great length of time and is riddled with uncertainties; much of the evidence has been lost. No one can go back to experience what life was like in an oral culture. Books in their myriad forms are now an integral part of our daily lives. We can only imagine what it must have been like to see the first alphabet, to feel the wonder of letters, to hold the instrument of writing in one's hand. Would the people of that earlier era have recognized that they were participating in a profound revolution, any more than we are able to grasp the implications of the digital age?

If this account of the transition from orality to script to print makes sense, it is because there is enough similarity between that earlier time and the current era to merit the application of our logic. Goody suggests that "logic, our 'logic', in the restricted sense of an instrument of analytic procedures...seemed to be a function of writing" (1977, p. 11). This story begins prior to such analytic procedures. Despite our best theorizing, we can only answer whether this particular account of the transition from orality to writing helps explain the curious world that now confronts us.

We are not engaged in the sort of scientific enterprise that can offer definitive claims of truth. If you are concerned with proof, I suggest that you presume the story I am about to tell is wrong. Then consider what explanatory powers it offers nevertheless. To the extent that this story provides a way of making sense of our current situation, it could stand as a myth of our time. It is not necessary to verify a myth; all that is required is that we listen in a way that allows the implications to become evident.

The term *orality* indicates a primary orality in which writing does not yet exist. According to Havelock in his work on Greek literacy, *The Muse Learns to Write* (1986), "the common use of the term 'writing' by specialists as applied to any and every form of symbolization without distinction has helped to blur the boundaries between primary orality, a distinct and separate condition of society, and its successors" (p. 65). Such a primary orality existed in early Greece and figures prominently in the writings of Plato. Havelock argues,

> The concept of selfhood and the soul, as now understood, arose at a historical point in time and was inspired by a technological change, as the inscribed language and thought and the person who spoke it became separated from each other, leading to a new focus on the personality of the speaker. (p. 120)

As we shall see, it was the invention of writing that made it possible to record one's thoughts and to reflect on that record, leading to a separation of speaker and inscribed language. As Havelock, Ong, Goody, and others have detailed, a community defined by the traditions of primary orality would have little opportunity for such a reflective distance.

Plato expressed apprehension about this inability to reflect critically on an oral narrative, arguing that such narratives could mesmerize the mind of the student. According to Havelock, Plato opposed the poets because they offered a kind of spell—an intimate identification with polymorphic vivid narrative, an appeal to the unconscious—that had to be challenged. Plato sought to replace poetry with Socratic dialogue, asking that the poet restate the points contained in his verse. Such a challenge required the poet to state the matter in prose: "to ask what it was saying amounted to a demand that it be said differently, non-poetically, non-rhythmically, and non-imagistically" (Havelock, 1963, p. 209). Plato was attacking not only the way in which information was communicated but also the poetic state of mind and, by implication, the consciousness that flowed from that state of mind. Plato argued for a reflective, thinking self; poetry prevented that, insofar as the poetic experience prevented autonomous, reflective awareness of what was being said.

Plato also opposed written forms of communication because he believed that writing would hinder one's perception of truth. As Johanna Drucker explains in *The Alphabetic Labyrinth* (1995), "Plato held all visual images in low esteem....For if apparent reality were a mere shadow of the Ideal, then any image imitating reality was at best yet another degree removed from the Ideal. The mark or letter, like any image, served a merely mechanical function and could not in any sense serve the highest function of human knowledge, to reveal Truth" (p. 60). Ong argues in *Orality and Literacy* (1982) that "writing, Plato has Socrates say in the *Phaedrus*, is inhuman, pretending to establish outside the mind what in reality can be only in the mind" (p. 79). Plato's polemic against writing and poetic verse was central to the establishment of his own system. Critical dialogue required a reflective awareness that was not possible when one was beguiled by images or verse.

It is difficult to guess whether or not Plato would view our profound reliance on writing as a confirmation of his worst fears, though he certainly might understand the perspective that the images of television are detrimental. By contrast, it might be very difficult for today's critics of television to agree with Plato's view that writing is also detrimental. Writing is no longer a new form of communication; it is the background that structures how one constructs a sentence, outlines an idea, and communicates everything from free verse to flow charts. Did Plato not understand the power of writing? Or did he understand all too well the power of a new medium to transform how people organize not only their ideas but also their communities?

In *The Interface Between the Written and the Oral* (1987), Goody states that changes in the means of communication subsequent to the adoption of speech may have important implications for the structure of ideas as well as for the structure of society. Long before Ong, McLuhan, Havelock, or Goody suspected that differing forms of communication alter one's experience of the world, Plato understood that differing forms of communication alter how one thinks, speaks, and remembers. As Havelock details in *Preface to Plato* (1963),

> Control over the style of a people's speech, however indirect, means control also over their thought. The two technologies of preserved communication known to man, namely the poetised style with its acoustic apparatus and the visual prosaic style with its visual and material apparatus, each within their respective domains control also the content of what is communicable. Under one set of conditions man arranges his experience in words in some one given way; under the second set of conditions he arranges the same experience differently in different words and with different syntax and perhaps as he does so the experience itself changes. (p. 142)

The gradual shift from an oral to a script culture constitutes perhaps the earliest transition from one dominant form of communication to another. Plato's argument that the right kind of discourse gave rise to an autonomous, thinking self was an idea constrained, in ways we can never know, by circumstance and history.

It is ironic that support for Havelock's views on orality and literacy must be carefully deduced from written texts, which provide the only remaining record of this period. Because we are born into a very different sort of experience, it is difficult to imagine this battle between the oral and written tradition. If poetry structures awareness differently than writing and if writing, in turn, gave rise to a new concept of the person, then what of the forms of communication today? How do current structures shape our awareness not only of information but of self and other?

Distinct Features of Orality

A study of early Greece proves that it would be quite impossible for an oral society to understand that there are very different ways of organizing information, because that awareness comes only when one has the opportunity to experience alternative forms. It is possible to characterize text as prose or poetry, to note that the organization of the words on the page contributes to how the ideas are read and experienced. Granted, there is a wealth of variations on this theme. There are poems that are more like paragraphs, and prose that is poetic in tone and form. There is *Finnegans Wake*, case law from the Supreme

Court, and everything in between. Poetry does not help when one is trying to explain the First Amendment; prose cannot adequately capture the poet's vision. But for the poet who had never experienced a diversity of written forms, such distinctions did not exist. As Havelock explains, when "Hesiod in effect asks, Who is the muse? What precisely does she do, and how does she do it?"(1963, p. 99), he has moved to define that content and purpose of poetry, which for the wholly oral minstrel had been unconscious.

Just as the poets of early Greece were unconscious of the ways in which spoken language structured their communication environment, our culture may be unaware of how different forms of communication structure our environment. In *Understanding Media* (1994), McLuhan argues that we, like the Greek poets, are mesmerized by our technologies. He believes that most theories of media ignore "the nature of the medium...in the true Narcissus style of one hypnotized by the amputation and extension of his own being in a new technological form" (p. 11). In our current milieu, poetry is an important form of literature, an artistic craft: we do not turn to the poet to explain scientific discoveries, social theories, or legal doctrines. For such debates we reserve rational discourse, prose, explicit paragraphs. We place our faith in those texts that most carefully reference earlier editions. But without such texts and earlier editions, without the concept of linear paragraphs, poetry served a very different purpose.

In ancient Greece, "poetry was not 'literature' but a political and social necessity. It was not an art form, nor a creation of the private imagination, but an oral encyclopedia maintained by cooperative effort on the part of the 'best Greek polities'" (Havelock, 1963, p. 125). Imagine an encyclopedia that must be remembered and recited, in order to preserve the knowledge of the community. The constraints of an oral encyclopedia would undoubtedly result in different ways of structuring and remembering information.

For those of us who cannot remember what is scheduled on our personal digital assistant for any given day, the very idea of remembering the entire history of the community is absurd. Imagine remembering who bought which parcel of land, who owes money to whom, remembering the customs and laws, the ages of the young and old, when elders died, which lands result in the best harvest, which herbs heal illness. There were no reference books to review or written records to check because an oral society, by definition, does not have such resources. Furthermore, people would be quite unaware that they lacked those resources, whereas we are all too aware that life would be untenable without them. In such a society poetry served as a form of instruction, a method of record keeping, and a way of preserving the history of the community.

The oral culture would inevitably structure knowledge very differently than we structure knowledge today. We test our knowledge of a subject by checking the facts, concepts, dates, and details. An investigation into a new subject will entail a systematic and logical analysis. In an oral society, knowledge operates by very different rules. Havelock has summarized the general character and content of knowledge in the oral culture under three separate aspects: First, "the data or the items without exception have to be stated as events in time" (1963, p. 180). Second, events are "remembered and frozen into the record as separate disjunct episodes each complete and satisfying in itself" (p. 180). Third, "these independent items are so worded as to retain a high content of visual suggestion" (p. 180). In other words, memory is better served if the story is situated as concrete events, occurring in a specific time, and communicated through highly visual elements.

Such an organizational structure precludes abstract or analytic forms. As Havelock explains, "Kantian imperatives and mathematical relationships and analytic statements of any kind are inexpressible and also unthinkable" (1963, p. 182). Plato demanded that "the Greek mind break with the poetic inheritance, the rhythmically memorised flow of imagery, and substitute the syntax of scientific discourse, whether the science be moral or physical" (p. 182). Plato was making a point about the medium: he was arguing that how one structures the information has a direct impact on what can and cannot be expressed.

Writing provided a new way of structuring one's thoughts, which gradually led to new forms of knowledge. Ways of thinking that we take for granted, such as systematic classification and analysis, would be quite difficult if our knowledge were stored only in oral form. The difficulty arises first with the need to accurately remember the oral recitation, and second from the constraints imposed by the spoken word. In order to remember and recite a lengthy story, the performer needed to devote significant attention to remembering the details, repeating the words over and over in order to be carried along by the rhythm and sound of the language.

It is not unlike the experience of saying one's prayers in church, allowing the repetition of the words to soothe the troubled mind. It is interesting to note that, for those raised in the Catholic faith, reading a prayer book is secondary to the experience of reciting the liturgy, the spoken mass echoing the orality of ancient Greek culture. It is impossible to remember large volumes of information without the metered rhythm, the repetition of sound to sustain and carry the memory forward. There are individuals in our literate culture who have needed to memorize large amounts of information. The PBS documentary *Return with Honor* tells the story of American prisoners of war in Vietnam. One American prisoner memorized all the names of his fellow captives in the prison camp. In order to recite those names, he understood that the rhythm and flow of the sounds must not be interrupted.

As Havelock points out, it is not just that the oral history is communicated as time-bound, discrete events; the formulas of storytelling required that the speaker pay more attention to remembering than to critically analyzing or categorizing what was said. There was, of necessity, no reflective distance between the speaker and the word. To draw again on the work of Havelock:

> As we pass from experience to experience, submitting our memories to the spell of the incantation, the whole experience becomes a kind of dream in which image succeeds image automatically without conscious control on our part, without a pause to reflect, to rearrange or to generalise, and without a chance to ask a question or raise a doubt, for this would at once interrupt and endanger the chain of association. (1963, p. 190)

Primary orality contributed to a view of the self that did not include the separate and autonomous perspective associated with the modern concept of individuality.

The belief in the inalienable rights of the individual arose in a modern world under very different historical circumstances. Such ideas are neither absolute nor enduring. For an oral culture it is the group, rather than the individual, that offers strength, protection, and continuity. The group bears witness to the customs of the community and shares the stories of the community. Those oral formulas would have made it difficult to disengage and critique the story. The necessity of remembering and the constraints of reciting long passages would have left little room for critical analysis, even if that possibility had occurred to the storyteller. Such reflection requires that one be able to separate oneself from the process of reciting. Prior to the invention of writing, this would have been a difficult task: "the whole community from minstrel and prince down to the peasant was attuned to the psychology of remembrance" (Havelock, 1963, p. 140). Oral cultures endured by focusing on tradition and custom, repeating the practices of the past in order to secure the future. Allegiance to the past and to the group encouraged conformity rather than autonomous reflection.

If, by chance, a creative freethinker had emerged in such a group, it would have been easy to silence that one voice through isolation. I am not suggesting that there would never have been disagreements in such a community but merely that the methods for settling those disagreements would have been vastly different from our methods today. Nor am I suggesting that the people of such a community were not aware of their own separateness from the group, as all people everywhere are able to distinguish between self and other. Rather, it would have been quite impossible to value individuality as we value it today, to applaud the freethinker, to privilege the individual over the group, or to grant a public forum for dissenting voices. Such was the power of the spoken word to bind a cohesive group to the task of remembering the past rather than inventing the future. The concept of individuality that serves as the foundation for an American way of life would have been inconceivable in a world where all memory had to be collectively recreated through the efforts of the group.

It is difficult to imagine a world where spoken language constitutes the only form of communication. Though such oral communities still exist today, they are hedged on all sides by a rapidly expanding communication revolution. The development of writing is identified as the first communication revolution, and it forever altered the social contract of the oral community. Some scholars trace the date and origin of the Greek alphabet to around 750 B.C. (Goody, 1987, p. 60). The development of printing contributed to a second revolution, transforming politics, religion, and economics. That revolution is traced to the invention of the Gutenberg printing press in 1450, though "rag paper and good ink were available in China by the year 200 A.D. or perhaps earlier. A few hundred years later, documents were being printed from carved wooden blocks; the oldest surviving book so printed was produced about 700 A.D." (Schramm, 1988, p. 117). Today, the third communication revolution—the hybrid mixed media of electronics—is transforming the social contract in yet other ways.

Whether one focuses on oral, script, or print cultures, there is a basic premise that must be recognized in all cases. This premise, widely accepted among scholars of communication and rarely recognized by a broader audience, advocates that a shift from one form of communication to another alters how information is organized and understood. Havelock argues that "this radical shift away from oralism...occurred in a proliferation of terms, for notions and thoughts and thinking, for knowledge and knowing, for understanding, investigating, research, inquiry" (1986, p. 115). According to Ong, "technologies are not mere exterior aids but also interior transformations of consciousness, and never more than when they affect the word" (1982, p. 82). Goody explains that "writing has a particular kind of internal influence since it changes not only the way we communicate but the nature of what we communicate, whether to others or to ourselves" (2000, p. 136). Such changes, in turn, can have a profound impact on how communities function. The formula of the oral epic constrained the story in specific ways. The reciting of such stories required a cohesive group of listeners who were willing to faithfully repeat the litany, discouraging independent thinking. Such was the spell of the spoken word. Just as the oral culture encouraged allegiance to the group, the emergence of printing created new forms of discourse and different social relationships.

From Script to Print

With the creation of a phonetic alphabet, communication was changed forever, and, in the process, the relationship of self and other was altered. Speech, script, and writing cannot be viewed as mere additions to one another. Script does not simply extend the range of speech. Print does not merely provide multiple copies of written manuscripts.

The experience of reading was profoundly visual in a way that spoken language could never be: "Writing had reconstituted the originally oral, spoken word in visual space. Print embedded the word in space more definitively" (Ong, 1982, p. 123). The spoken word resonates, surrounds, and draws the listener into the speaker's experience, creating a shared sound space that is both acoustic and fleeting. The spoken word draws people together, whereas print isolates. The text requires that the eye follow the uniform line across the smooth and silent page. The solitary reader imagines the voice of the author, but it is an imagined voice. McLuhan argued that, as a culture becomes visually dominant, the sensory balance is altered, and the consequence is a form of dissociation. McLuhan, Ong, and others recognize the importance of this "sudden breach between the auditory and the visual experience" (McLuhan, 1994, p. 84). As Ong explains:

> Spoken words are always modifications of a total situation which is more than verbal. They never occur alone, in a context simply of words. Yet words are alone in a text. Moreover, in composing a text, in 'writing' something, the one producing the written utterance is also alone. (1982, p. 101)

Print allowed for new ways of classifying and recording experience, but it also allowed for separation from the lived world and from the web of kin and clan. The alphabet encouraged new ways of structuring knowledge. One's thoughts could be spelled out, edited, rearranged and reviewed. That level of analysis was not possible prior to written forms.

A cursory understanding of these distinctions can be gained by examining one's own practices, observing how one responds to face-to-face conversation, and noting what is different about the activity of reading a book. The words on the page create a solitary space that invites reflection. The eye follows the paragraph, the line of reasoning, the story line, the linear progression of the text. Print isolates not only the reader but also the eye. Reading is a profoundly visual activity; there is no intonation, no breath of the living word. The printed page holds the world at a distance.

The handwritten manuscript served as a transition from oral to print cultures, displaying some features of both orality and text: "The physical properties of early writing materials encouraged the continuance of scribal culture" (Ong, 1982, p. 94). A handwritten manuscript took years to produce, there were a limited number of copies, and few people could read. For all of these reasons, the manuscript was often read aloud, and in this way script continued for a time as part of the oral culture.

With the invention of the printing press, it was possible to create multiple copies. As those copies were circulated, reading became more widespread. The conformity of the oral community was usurped as dissenters found a forum for their views and beliefs. Martin Luther's theses tacked to the doors of churches stand as an often-cited example of the power of print to spread a dissenting view. As Eisenstein details in *The Printing Revolution in Early Modern Europe* (1983), when Martin Luther was called by the pope to exonerate himself, he responded that it was a mystery how his Ninety-Five Theses "were spread to so many places" (p. 148). In an oral culture, speech is shared only as far as the many voices of that community can reach. Because speech is ephemeral, there can be no reference to a fixed and permanent copy of what was said. In a print culture, there are multiple permanent copies, and it is difficult to control who reads those copies: hence the spread of new ideas, often considered heresy by the church or seditious libel by the government. These powerful institutions were quick to feel the threat of this new form of communication. Censorship followed as powerful institutions attempted to restrict the flow of information.

The alphabet distances the reader from the world in yet another way, interjecting a detailed and phonetic code that stands for, or represents, the world. The phonemes that constitute the word apple bear no resemblance to the round, red, smooth object that rests neatly in the palm of my hand: "The phonetic alphabet invented by ancient Semites and perfected by ancient Greeks, is by far the most adaptable of all writing systems in reducing sound to visible form" (Ong, 1982, pp. 91–92). The printed text becomes the intermediary of the world, insofar as the fixed words on the page represent that world. McLuhan and others argue that the phonetic alphabet altered one's relationship to the world and, in doing so, exacted a price of which few were aware. Literate man, he argues, existed largely in a world of disassociation and fragmentation. Experience was parsed into phonemes, syntax, categories and subcategories.

In order to grasp the way in which the text mediates experience, one need only compare the constraints of a well-structured paragraph, laid out in neat lines on the page, to the patterns of ordinary speech. The paragraph creates a structure that is very different from the intimate give and take of spoken language. If one considers the extensive schooling required in order to write clear paragraphs, it becomes obvious that this process is quite different from talking. Print extends one's reach, providing the opportunity to pursue and develop abstract ideas that would be impossible without a way of recording one's thoughts.

Ong writes of the alienation print created, of the ways in which one's experience of self was altered to include a sense of absence or of loss (1982, p. 82). All serious students have experienced that loss as they realize that their education extracts a price: that, figuratively speaking, they cannot go home again. Though writing is not the sole cause of all changes in consciousness, Ong understands that both orality and writing are necessary for the evolution of consciousness. He argues that "the use of a technology can enrich the human psyche, enlarge the human spirit, intensify its interior life" (p. 83). But the corresponding loss entails a heightened awareness of one's separateness.

The pursuit of the text, of abstract ideas, of scholarly achievements entails a distancing from both the lived experience of the world and the felt sympathies of the group. As Ong explains:

> By removing words from the world of sound where they had first had their origin in active human interchange and relegating them definitively to visual surface, and by otherwise exploiting visual space for the management of knowledge, print encouraged human beings to think of their own interior conscious and unconscious resources as more and more thing-like, impersonal and religiously neutral. Print encouraged the mind to sense that its possessions were held in some sort of inert mental space. (1982, pp. 131–132)

The implication was that one's thoughts exist like a kind of text captured in the uniform letters on the flat page, that it might be possible to fully detail one's thoughts. To understand the technology of print, one must grasp the consequences for the individual reader, who is isolated in curious ways from both self and other. The paradox is that the reader, though isolated, can also draw closer to the experiences of the author, delving into the personal thoughts of another human being as never before.

The creation of books led quite naturally to the idea of authorship, of creative expression, originality and ownership. As Ong explains, "print culture gave birth to the romantic notions of 'originality' and 'creativity,' which set apart an individual work from other works" (1982, p. 133). As printers, writers, and editors struggled to define their respective roles, the concept of authorship emerged as a factor of the marketplace. According to Eisenstein, "the repertoire of roles undertaken by early printers seems so large as to be almost inconceivable. A master printer himself might serve not only as publisher and bookseller, but also as indexer-abridger-translator-lexicographer-chronicler" (1983, p. 61). In an oral culture, the history of the community was shared by all and owned by all. With the development of the printing press, it became possible to identify individual authors. The economic implications and the resulting legal ramifications are of less immediate interest than the way in which the concept of ownership designates a growing awareness that ideas can belong to a particular individual.

Print contributed in significant ways to the emergence of individualism. The alienation of which Ong writes indicates a separation from the group. The cohesive bonds of the oral community are broken, and what gradually emerges is a sense of the human as a separate and autonomous individual. Because individualism, and the rights thereby assigned to the person, is fundamental to American democracy, it is difficult to realize that this concept is of fairly recent origin. The concept of individualism grew out of the context of the counter-revolutionary critique of the Enlightenment and can be linked to an emerging recognition of the autonomy of the person; to social, religious, and economic changes that swept the European continent with the rise of the modern world; and to new forms of government and the development of psychology (Wiener, 1973). Despite this wide range of precipitating factors, there are a number of scholars who argue that the emergence of individualism can also be linked to the development of writing and printing.

In summary, printing encouraged new ways of organizing information, challenged the authority of dominant institutions, increased the diffusion of printed material, and encouraged the solitary pursuit of new ideas. Each of these points facilitated an awareness of the person as autonomous from the community. Individuation is enhanced to the extent that the person has the ability to reflect on his thoughts, to analyze and critically evaluate his position on various issues. For that work, the printed text offered distinct advantages. As Eisenstein details, the ready availability of printed material challenged the authority of church and state in a way that script could not have. With script there were a limited number of manuscripts, entrusted to the church or to wealthy patrons. Few people could read, and fewer yet had access to a handwritten manuscript. Knowledge was controlled by those with the wealth and power to collect the few available manuscripts. With print, this monopoly was broken as identical texts flooded Europe.

As people learned to read and write, they based their study of the Bible on their own interpretations rather than the word of the Church. They developed theories about everything, from dress and manners to politics and law. As Eisenstein explains, the increasing diffusion of a wide range of materials, including texts, catalogs, indexes, and manuals, provided an accessible forum for people to compare their customs, habits, and thoughts to the customs, habits, and thoughts of others. This ability to compare one's social practices to a wider group, independently of one's immediate social circle, encouraged new attitudes about roles and identities in the social domain. The activity of creating a record of one's thoughts, and the opportunity to read the thoughts of others, encouraged an awareness of one's inner life that was simply not possible without the availability of the printed word. Thoughts that would not have been publicly expressed, situations that could not be described in polite company, could now be detailed on the printed page. The solitary activity of reading could encourage both reader and writer to reflect on inner experience in a way that was not possible prior to print.

A New View of the Person

Printing also contributed to a new concept of privacy, which may have begun simply as the idea of a personal space that was separate from the public domain. As M. Ethan Katsh explains in *The Electronic Media and the Transformation of Law* (1989), "privacy, like copyright and obscenity, had no direct legal ancestor in the preprint era. As a legal concept, privacy is something new, a field of the law whose origin is linked to printing" (p. 189). The first legal discussion of privacy surfaces in an article by Samuel Warren and Louis Brandeis, published in the *Harvard Law Review* in 1890. The two young Boston lawyers propose "the creation of something novel, a legal right of privacy" (Katsh, 1989, p. 190). The concepts of privacy and individualism are deeply related, in terms of both origins and legal implications. If one of these concepts is altered by time and circumstance, it is probable that attitudes about both concepts will be called into question.

In *Privacy and Print* (1999), Cecile M. Jagodzinski writes that, for the early moderns, privacy carried the suggestion of individual, rather than public, interest. In early Greece, privacy carried negative overtones, implying lack of full participation in the social order. According to Barrington Moore, the Greek term for 'private' was *idios*. "The noun form of *idios* is *idiotes*, from which comes the English word 'idiot.' Its main meaning is a private person or individual, or one in a private station as opposed to one holding public office or taking an active part in public affairs" (Moore, 1984, p. 82).

The term *individualism* can be used in a variety of ways to describe religious, political, economic, philosophical, and social implications. Robert N. Bellah et al.'s *Habits of the Heart* (1985) details some of the permutations of this concept that have played out across the American landscape, from the communitarian to the rugged individualist. In *The Saturated Self* (1991), Kenneth J. Gergen argues that identity is a social creation, a response to the diverse roles entailed in ordinary life. Here individualism is nothing more than a social constraint. By contrast, the American transcendentalists advocated that it was the solitary self, removed from social constraints, that offered the clearest path to one's true self.

The concept of individualism as we know it began in a particular time, has evolved in response to changing circumstances, and at some future point may, perhaps, cease to be relevant. Because American political, legal, and economic systems are based on shared concepts of the autonomy and privacy of the individual, it is difficult to imagine what life might be like without these organizing concepts. Katsh argues that, "at the heart of the modern liberal legal system is a 'view that society is constituted of autonomous, equal units, namely separate individuals and that such individuals are more important, ultimately, than any larger constituent group'" (1989, p. 239). Inevitably, new forms of communication reconfigure how knowledge is organized and understood. Ong, Havelock, Goody, and others argue that such shifts in the organization of knowledge have an impact on the social domain.

As a consequence of changes in communication technology, prevailing concepts of privacy are under attack. Current challenges to privacy highlight the need to protect personal information such as credit histories, medical histories, and financial information from public scrutiny. What is not yet evident is the extent to which the struggle to clarify new definitions of privacy mirrors comparable struggles with the concept of individuality. What is needed is a richer awareness of the ways in which these terms are related.

Comparisons can be drawn between the kinds of changes the transition from orality to print created and those created by the transition from print to electronic communications. An understanding of the changes evoked by the transition to writing and print prepares the way for understanding how current communication technologies are altering the role of the self in the social world. While there are parallels that can be drawn from those earlier revolutions, there are also distinct differences between the current revolution and earlier changes in the modes of communication. First and foremost, when considering the transition from orality to print, the lengthy passage of time provides a perspective. When considering the events of the late twentieth century, historical analysis has not yet provided an organizing framework.

The shift from an oral to a print culture has been characterized as a shift from the ear to the eye. The ear surrounds, whereas the eye distances; the ear is unidirectional, whereas the eye isolates. Reading is a visual activity. Our current mix of hybrid media does not easily reduce to discussions of eye versus ear. Technology now extends the senses in unprecedented ways. The visual effects of print are nothing like the visual effects of television. The experience of the screen is visual, auditory, and kinesthetic, insofar as the image incorporates vision, sound, and movement. The surround sound of a wide screen can intensify all the senses.

One avenue for understanding the current mix of media is to ask how the sensorium is altered. We live in a furious cacophony of sound and sight and movement. A visit to a local mall confirms the dizzy mix of stimuli that constitute our public spaces. Without even addressing the effects of communication technologies, one notices cascading escalators, gyrating amusement rides, and the vertigo of glass elevators. This highly visual experience is intrinsically related to the visceral experience of motion. The altering of the sensorium appears, at first glance, far more intense and problematic than the transition from ear to eye. Would Plato rail against this new age as he railed against the poets? Would he argue that this mix of vision, sound, and movement renders the student less capable of critical reflection? Certainly that is the argument that is being played out today, with reference to television rather than the shopping mall. But television is not the issue, if it ever was. The effects of television are matched on all sides by CDs, video games, digital photography, word processors, web pages, audiotapes, advertising as well as the shopping mall, which is also a form of communication.

Consider another element of the transition from orality to print. As explained earlier in this section, print allowed for new ways of organizing information. The oral tradition required that all information be communicated in a concrete and time-bound fashion. By providing a fixed record, print allowed for new systems of classification and new forms of discourse. One need only compare the structure of a linear paragraph to the structure of an epic poem to recognize the difference. Does the current mix of hybrid media also provide new ways of classifying knowledge and hence new ways of thinking? There are scholars who argue that this is the case. Writing about the current communication revolution, Richard A. Lanham argues that "at the center of this repositioning in the human sensorium stands a major readjustment of the alphabet/image ratio in ordinary communication" (1993, p. 125). He offers an argument for broadening the way we think about human reason.

If printing contributed to a sense of separation from kin and clan, to the rise of individualism, what can be said of the communication revolution today? In America we have a limited set of terms to explain the self in the social world. We refer to private and public life, to the rights of the individual, to the needs of the group. We distinguish between inner and outer experience, designating inner experience as belonging to the private person. These concepts may no longer adequately address key features of the current milieu. If earlier communication revolutions altered the role of the self, it is likely that the changes evidenced today will also have a significant impact on our shared understanding of the self in the social world. What is needed is a thorough account of how today's communication revolution is altering shared concepts of the self.

How, then, does this account of the transition from orality to literacy provide a framework with which to understand the current information revolution? Scholars have implied such a relationship through the use of the term *secondary orality*, which indicates a communication environment that includes not only books but telephones, radio, and television (Ong, 1971, 1982; Havelock, 1986). Ong observes that, "with telephone, radio, television and various kinds of sound tape, electronic technology has brought us into the age of 'secondary orality.' This new orality has striking resemblances to the old in its participatory mystique, its fostering of a communal sense, its concentration on the present moment, and even its use of formulas" (p. 136). The term *secondary orality* is not meant to imply a straightforward return to primary orality: "What had happened was not a reversion to a primeval past but a forced marriage, or remarriage, between the resources of the written word and of the spoken" (Havelock, 1986, p. 33). McLuhan discusses the concept of secondary orality in terms of retribalization. He notes that television and other forms of mass media reintroduced an awareness of the group that had been absent in a print-dominated culture.

Both Ong and McLuhan understood that electronic forms of communication in the twentieth century were undermining the dominance of print. But it is an open question whether the secondary orality of radio and television is anything like the primary orality that existed prior to the invention of writing. Both television and radio exist in a highly literate milieu, an environment that print influences in subtle and profound ways. Hence the term *post-literate*, which indicates an experience that both incorporates literacy and moves beyond it. The question is, what does it mean to move beyond literacy? Multimedia do not merely constitute an addition to print. Rather, digital images, video, and sound create new forms of communication. Hybrid media are much more than fancy books. Many forms of media in the digital age require new literacies that are unrelated to print. The transition from orality to print has been superseded by a third transition that moves beyond print. The term *post-literacy*, like *secondary orality*, draws on an earlier experience to make sense of the present. The challenge is to understand how newer forms contribute to an experience that neither mirrors earlier forms of orality nor mimics the constraints of literacy.

Secondary orality is a useful concept for indicating a certain resemblance between primary orality and the features of broadcast media such as radio and television. It is not clear that the concept provides an adequate explanation regarding key features of today's hybrid media. Primary orality relied on formulaic rhythm and repetition, on concrete events located within a specific time and place. Primary orality privileged the group in a way that excluded individualistic perspectives. Primary orality captured the experience of a community for which the voice was the dominant form of communication. Few, if any, of these details are characteristic of hybrid media. The screen—whether DVD, streamed on the web, or broadcast live—presents a complex mix of sensory data that cannot be reduced to the eye versus the ear.

The relationship of listener to speaker is also complex. The mass audience has been fragmented with interactive media and narrow casting. Ong, McLuhan, and Havelock are correct that the mass media fostered a group sense that was different from group awareness in print-dominated societies. Electronic media have further extended and enhanced awareness of groups in unpredictable ways. The concept of secondary orality does not adequately address the complex features of group awareness facilitated by new media.

In addition, key features of primary orality are also absent. Hybrid media do not constrain autonomy and encourage conformity as was the case with primary orality. Quite the contrary, it appears that individuality is enhanced as more people program personal web pages, record and communicate personal moments with digital cameras, scan, save, and file images from a growing abundance of visual forms. Hybrid media do not require that ideas be communicated in concrete details, located in a specific time and place as with the oral tradition. Just the opposite is occurring. It is now possible to visualize ideas as never before, in a variety of models, many of which can be electronically manipulated by the viewer.

If the story of the transition from orality to print is a myth for our time, it is because of the similarities between that communication revolution and the current one. The transition from orality to print transformed how information was stored, accessed, and shared; new ways of classifying and storing knowledge altered the structure of communities and contributed to an evolving perspective on the autonomy of the self. Each of these points can be applied to the information revolution that is currently reshaping nations, politics, economics, and communities. In both cases, there is a complex interplay between emerging forms of communication and the resulting shifts in knowledge and social structures. Communication technologies, whether print, monitor, or virtual reality, contribute to how one experiences and conceptualizes the world. Those new ways of conceptualizing the world have consequences for the social domain.

Rather than relying on the concept of secondary orality to explain what is now occurring with new media, the distinction of ear versus eye can be used as the phenomenological starting point for an investigation of the communication matrix. The work of McLuhan, Havelock, Goody, and Ong provides a framework that establishes the connections between shifts in the human sensorium and the social milieu. What is needed is an understanding of how one's perception of the world has been altered. The transition from orality to print can help us to better understand the transition from print to electronic forms of communication. The challenge is to understand how shifts in the human sensorium alter social structures, which, in turn, lead to changes in the concept of the self. A phenomenological investigation of the screen will reveal the ways in which the social construction of the self is being transformed.

Chapter Three: Altering the Constraints

While the emergence of literacy may have been tumultuous, history reduces that transition to a number of progressive steps with an understandable conclusion. It is beyond our human powers to guess what confusions reigned during the centuries it took to make such a transition. The passage of time softens contradictions and ambiguities. It is easier to reconfigure the past through a narrative than to understand the confusions of the present. The events of the twentieth century are too recent. Faced with unprecedented changes, it is difficult to discern an organizing theme. It took centuries to make the transition from orality to script and centuries more to create movable type. In retrospect, it appears that the consequences of writing unfolded in slow motion against a backdrop of social, economic, and political change. In the current era, communication technologies follow each other in rapid succession; we are unable to see a gradual evolution of forms.

Radio, telephone, film, and video signaled only the beginning, as new communication technologies continually provided different formats for sharing knowledge. The terms *new media, hybrid media,* and *mixed media* allude to the ways in which the environment has become increasingly mediated. It seems as if Pandora's box has let loose myriad fanciful forms to confound the senses. There is the wish for a myth of our time. The desire is to understand the present from the vantage point of another era, to discern by insight the key features of a landscape caught in a profound transformation. In this pivotal age one senses the possibility of myths in the making, though it is difficult to know what to pay attention to and what to ignore. It appears that we are confused and amused by technological inventions that prove to be anything but toys.

The record of the transition from orality to literacy is a myth for our time because that record reveals, in mysterious ways, the emerging story of this era. Insofar as the printed page emphasized the isolation of the eye, this framework provided a way to characterize the difference between speech and print. Our current era cannot be captured by the formula of eye versus ear. We are all eyes and ears and more, wired in every conceivable direction for sound, image, text, color, and movement. We lack a framework to explain our hypermedia environment. Nevertheless, the story of the transition from orality to print identifies key elements that can help us better understand the present.

In each case, mediation has led to new forms of knowledge. Writing provided new ways to record experience, freeing the mind from the arduous process of remembering. Print gradually contributed to a growing awareness of the self as a separate, thinking subject. While the current revolution in communication technology may alter collective memory in other ways, the more important point is that the digital age alters collective awareness of space and time. Hybrid media create a new space conception, with profound consequences for human relationships.

Scholars have relied on the distinction of eye versus ear to explain not only orality and literacy but the difference between print and electronic forms. John Fiske draws on oral and literate modes to explain television; his point is that "the formal characteristics of television are essentially those of oral rather than literate modes of communication" (1987, p. 106). Suggesting that "our culture is one that validates the literary, or rather the literate, and consequently devalues the oral" (p. 105), Fiske classifies orality as dynamic, mosaic, concrete, social, episodic. In turn, he classifies literate modes as sequential, linear, abstract, individual, and narrative. Though these terms reflect McLuhan's perspective in *Understanding Media*, a careful analysis reveals that they may not successfully clarify how print and electronic cultures differ.

What is it that makes the screen mosaic? What exactly makes print linear? McLuhan's point is that each medium organizes information in a different way. My point is that the general use of such terms does not adequately capture the fundamental differences between the book and the screen. There is nothing inherent in the medium of television that insures dynamic, mosaic, social forms, just as there is nothing inherent in the print medium that guarantees sequential, linear, individual forms. Depending on how these terms are applied, they can characterize either medium.

One may watch television as a solitary individual in the same way that one reads a book alone. The content of television may be borrowed from film and based on a linear, sequential narrative. The book, on the other hand, may detail poetry, prose, or anything in between with varying degrees of linearity or nonlinearity. Furthermore, to the extent that a book is read by many, it will become a shared experience not unlike the shared experience of television.

Yet it seems there is something to the dichotomy of orality and literacy that might help explain how our highly mediated world has moved beyond the dominance of print. Paul Levinson explicates McLuhan's view, arguing, "the world that comes to us after the alphabet, in the form mostly of television, which is also musical, mythic, immersive, and, unlike the book and the newspaper, lacks perspective or distance from its subjects" (1999, p. 46). Again, anyone who has read a good newspaper knows it can be immersive. In fact, as McLuhan points out, newspapers have more in common with television than other kinds of printed material, given the mosaic layout of newsprint. Good books can engage the reader in a way that dissolves all distance and perspective. So what exactly is the difference between a print-dominant culture and an electronic culture?

It is not clear whether the terms *mythic* and *immersive* point to the nature of the viewer's experience, highlight the nature of the medium, refer to specific content, or reference all three. Ong's use of the term *secondary orality* is far more understandable, insofar as he argues that mass media provided access to large groups in a way that was not possible prior to the invention of radio and television. But when the concept of orality is invoked in a diffuse and general way, to imply nonlinear, configural, mythic, and immersive forms, it is not helpful. McLuhan has something important to say about the difference between print and electronic forms. In order to grasp his point, we must do more than invoke his terms. We must understand how the screen differs from the book.

The Screen as Portal

Half a century ago, there were few screens available. Television was in its experimental stages, and some wondered how one might use such a strange invention. Computer monitors had not yet been invented and video games did not exist. Today, most people are surrounded by screens at home and work. We fail to notice that at airports, banks, offices, homes, in cars, on the street, and in restaurants, the screen glows with images, sounds, and text. Current use of the screen has not yet evolved to provide the surveillance predicted in George Orwell's text, *1984*, but the merging of computer and television makes that a possibility.

McLuhan devotes one chapter each to film and television in *Understanding Media*, the only references in that text to a screen. But in 1964, film, and especially television with its limited channels and poorly defined visuals, was very different from the digital, high- definition, stereoscopic effects possible with today's screens. Each iteration of the screen enhances and enlarges the presentation of a world that is both intimately familiar and yet different from what can be experienced without the benefit of technological intervention. The images of the screen create an existential context in which temporal sequencing is altered and space is experienced as dynamic. It is the actual event, televised live, that most captivates. There is fascination verging on disbelief: we never realized the world looked like that. And if the camera could only explain itself, we would know that, prior to moving pictures, the world never looked like that at all.

It is not just that the image moves but that space and time are altered. The duration, the felt sense of space, time, and motion, is unlike anything experienced in ordinary life and yet feels real. The viewer fails to notice that such chronological and spatial coercions cannot be experienced without the benefit of mediation. I am not referring to avant-garde theater but to the pervasive use of video, film, and computer graphics. The electronic media alter space and time in a way that print never can. The viewer is drawn empathetically into that felt sense of motion in a way that is simply impossible with print.

The screen reveals what print is incapable of spelling out on the flat page. That joining of sound, space, and motion contributes to the verisimilitude of an altered duration that transforms one's experience of space and time. It might be that children growing up with television and computers can play out their ideas in slow motion or reverse and think in images as well as words. Whether one is watching a film, playing a video game, viewing the evening news, or exploring the visual structure of biological forms, the moving image provides a wealth of information that is not reducible to that which can be communicated through linguistic terms, whether spoken or printed.

When Ong writes about the transition from orality to literacy in terms of the eye and the ear, he is emphasizing the way in which the image of the printed word is detached from the sensual immediacy of the lived world. As McLuhan points out, the isolation of the eye altered the ratio of the senses. The emergence of a visually dominant culture followed. Though both printed words and pictures may be referred to as images, though both forms may draw on the eye, print operates quite differently from the screen.

Consider the difference between printed words and words on the screen. Printed words are fixed, do not float, cannot be enlarged, reduced, or hyperlinked. As a consequence, one's reading practices differ significantly for hard copy and electronic text. Though text dominates the Internet at this point, that will change as digital technology makes it increasingly easier to communicate graphics and video. When all information can be digitized, there is little technical difference between communicating words and images. Ong's study of the word helps one to understand the profound differences between spoken and written language. Through centuries of acculturation the Western world became print dominant, as the book shaped discourse and scholarly practices. New forms of media that utilize images in a distinctly different way now challenge the assumptions of that print-dominant culture. Video, film, and computer imaging do not isolate the eye, do not organize information according to the practices of print, do not separate action and object.

The screen is a portal, a gateway allowing for a kind of transport. One must understand the nature of that portal to grasp the essential features of the screen. Science fiction depicts the portal as a means of time travel. A character is in one century, and then, with a flash of light, she finds herself in another century. Anyone who has traversed a fascinating print narrative can identify with that sense of being transported. Reading is valued because it can introduce the reader to worlds that would otherwise be closed.

Yet the experience of reading and the experience of viewing moving images on a screen are fundamentally different. The layman may think of reading a book as an activity that requires effort, whereas viewing the screen is thought to be a more passive process. McLuhan's work challenges such neat assumptions by pointing out the complexity of watching the screen. Though his perspective was visionary, McLuhan's use of the terms *hot* and *cool, fragmentation* and *depth involvement* did not detail the fundamental differences between reading and viewing. Critics may debate whether the screen is mesmerizing, hypnotizing, or desensitizing, but such discussions miss the point.

The screen as a portal grants entry to a world where different rules of engagement apply and consequences are not fully understood. The screen contributes to the creation of a syntax based on movement in space: a space that is visual, kinesthetic, and acoustic. Digital forms heighten and intensify the expressive components of that syntax. What is needed is a level of analysis that captures the logic of movement and duration. The practice of writing and reading created new forms of public and private experience. In turn, the omnipresent screen serves as a portal that reconfigures public and private experience in ways that are fundamentally different from print culture.

The Figure-ground Gestalt

The screen serves as portal in that it alters one's perception of foreground and background. A gestalt of foreground and background is necessary in order to render perception meaningful. According to Torben Grodal, "in gestaltist language, grouping means that the 'whole' is more than the sum of its parts" (1997, p. 65). As one foregrounds or focuses on certain elements, other elements slip imperceptibly into the background: "One of the most elemental structural forces operating within the screen is the *figure-ground princi-ple*....Depending on what you determine the figure to be, the ground will change accordingly" (Zettl, 1999, p. 99). The figure-ground prin-ciple applies to both spoken words and visual experience. If I attend to every individual word in your utterance, I will be unable to under-stand the sentence as a meaningful whole. In order for the gestalt to emerge, something must be given up; perception demands that atten-tion be drawn away from one element, as it is drawn toward another element. The background is always there, providing the frame or con-text that supports the foreground. As long as one focuses on the fore-ground, there does not appear to be any tension. The proposition, the point, the image, the message is clear. But let the perceptual bounda-ries shift, and ambiguity emerges. The eye struggles to resolve the vis-ual image; the ear struggles to identify the spoken words.

Disorientation results when one cannot distinguish between back-ground and foreground. Each medium of communication establishes the figure-ground dichotomy in different ways. The arts explore this dichotomy, and creativity depends on it. To make a statement about something, whether in words or images, dance or music, it is neces-sary to foreground certain elements in a way that pushes other ele-ments into the background. What constitutes the background for the screen is very different from what constitutes the background for print. Understanding the ways in which different media constitute the figure-ground gestalt helps to illustrate what is unique about the screen.

Each technology foregrounds different elements: "It's not simply a question of communicating a single idea in different ways but that a given idea or insight belongs primarily, though not exclusively, to one medium, and can be gained or communicated best through that me-dium" (Carpenter, 1960, pp. 166–167). Radio foregrounds the spoken voice. As McLuhan explains in *Understanding Media*, "if we sit and talk in a dark room, words suddenly acquire new meanings and differ-ent textures....All those gestural qualities that the printed page strips from language come back in the dark, and on the radio" (p. 303). The human voice becomes the unifying theme that orients the ambient background of music and sound effects. With its heightened orality, radio communicates elements that cannot be conveyed through other media. Radio relies on the ear, and so it carefully situates foreground and background according to auditory cues. As Ong points out, radio is reminiscent of an earlier oral culture.

But compare radio to the telephone, which also relies on the ear, and it becomes evident that these two technologies foreground different elements. The telephone can be both personal and impersonal in a way that is nothing like radio, in that it foregrounds not the singular voice but an interpersonal conversation. Personal, because the voice of the other in one's ear confirms and communicates through intonation. Impersonal, because the other speaks at a distance, is invisible, escapes one's glance. Radio can provide many different kinds of programming. The telephone allows for many different kinds of conversations, some perfunctory and others more intimate. Nevertheless, with radio, the human voice of the announcer provides the continuity that defines the on-air experience for the listening audience. With telephone, the voices speak person to person, and it is the ongoing conversation that provides the continuity.

While it is correct that radio and telephone both utilize the ear, it may be more important to understand what each technology foregrounds. If a given insight belongs primarily to one medium, as Carpenter suggests, then understanding the figure-ground of a particular medium can help clarify the nature of that insight. How is it, in a particular case, that this ground gives rise to that figure? What are the implications of figure and ground for the communication matrix? That the telephone foregrounds conversation tells only part of the story. The experience of the other's voice speaking into one's ear, of listening for intonation to fill the gap of what cannot be seen, the sound of an almost inaudible breath as the other pauses: these are the details of a background that situates and confirms the conversation. Contrast this type of oral/aural conversation with a conversation recorded in print.

A conversation on the telephone is nothing like a conversation recorded in print. It is understood that print cannot capture nonverbal cues, such as intonation. Print draws on the eye to the exclusion of the ear, but that point does not explain what print foregrounds. As McLuhan, Ong, Havelock, and others understood, print foregrounds space.

his emphasis on space can be explored in terms of layout and type-
ice as evidenced by the creation of the page. This point can also be
iscussed in terms of narrative and closure, as in Ong's work. Or, this
oint can be understood in terms of linear versus configural forms as
videnced in McLuhan's work.

The book is not a mere container, not an envelope that holds one's
leas like a letter. Rather, print organizes those ideas according to the
ructure of the book, and that structure is based on a concept of space
iat creates a clear progression from beginning to end. The concern
ith organizational features of print, with the structure of narrative,
ie layout of pages, fonts, and chapters, all highlight the ways in
hich the book emphasizes space. The background can be identified
y asking what is taken for granted. In order for the book to have de-
eloped as it did, it was necessary to have a phonetic alphabet. In our
estern, print-dominant culture, it was the phonetic alphabet set to
pe that contributed to the structure and space of the book. Other
riting systems and other technologies can create a different style of
ook, but not the book as we know it.

Thanks to the work of Ong, Goody, Eisenstein, and others, there is
well-developed scholarly tradition that thoroughly investigates the
ifluences of the alphabet, the printing press, and the resulting spatial
nplications. Song does not foreground space. The lyrics of song alter
ie figure-ground relationship in ways that are quite different from
rint. A detailed analysis reveals the ways in which print provides a
)ace for one's ideas that is quite different from thoughts expressed in
:her media. When two friends have a face-to-face conversation, their
:tention will be drawn to more than just the words of the conversa-
on, unlike the telephone conversation, which foregrounds the vocal
ements. Each technology shifts the figure-ground to emphasize cer-
in elements over others. The dichotomy of figure-ground can reveal
ie unique influence of different media.

Our primary engagement with the world is through sight, sound,
)uch, movement, and, to a lesser degree, through smell and taste.
ne's non-mediated experience of the world integrates sight, sound,
id touch in a seamless web. Auditory processes direct the gaze, as vi-
on substantiates one's sense of kinesthetic balance, which confirms
irection and distance. It was Merleau-Ponty's task to interrogate the

ways in which bodily awareness of touch, sight, and hearing constitutes meaningful experience. The figure-ground for Merleau-Ponty is embodied by the hand that touches the world and, in turn, is touched by that world. To be both a thing in the world—an object—and one who looks at things—a subject—necessitates a figure-ground gestalt that confirms one's existence through reciprocal ways of being. Perception, for Merleau-Ponty, is a form of interrogation; the physical body provides the field for that interrogation.

Mediation translates one's primary experience of the world into other modalities, altering the figure-ground gestalt in the process. Communication technologies can both extend one's reach and obscure one's relationship to that phenomenological world, confounding the relationship of object and subject, perceiver and perceived. For example, though both speech and print involve language, print alters consciousness in unforeseen ways. Or consider that one's ordinary view of the horizon is quite different from the horizon portrayed through video, though both involve vision. If I understand that the telephone foregrounds verbal conversation, and if I, in turn, understand what has been minimized as background, it becomes possible to grasp how that technology obscures my relationship, in this case, to the immediate other.

Consider the difference between writing a letter and sending an e-mail. Both of these forms rely on text, and yet the figure-ground for each activity is quite different. Presume that the letter is handwritten. Remember the way in which the hand catches and spells out the mind's thoughts, the way in which the hand pauses to find the next thought, scratches out the last sentence and begins again. With e-mail it is also possible to edit, delete, or pause as one catches the next thought, yet the process of writing an e-mail feels nothing like writing a letter by hand. The kinesthetic activity of writing in script foregrounds the motion of the hand, which provides a pace that, in turn, influences the unfolding of one's thoughts. It is reflective in a way that is not quite possible with e-mail, by virtue of the hand's rhythm as it moves "through" the thought.

Think now of the ten fingers floating on the keyboard, the luminescent screen that displays the letters in perfect synchrony, the familiar sound of the keystroke. The movement of the fingers is nothing like the hand's curvilinear script. It is reasonable that these two activities feel very different, because the technologies of pen and keyboard are profoundly different. What one might casually refer to as a feeling, in fact, impacts how one thinks when engaged in either activity. The e-mail foregrounds the images of the monitor, as the glowing screen calls attention to the letters, which link to the fingertips. This activity is much faster than writing out a message longhand. The display is vastly different from the flat, smooth page laid upon the table. The monitor "monitors" the writer of the e-mail, who keystrokes the message without rereading, in a hurry without quite knowing why. The incessant rhythm of the ten fingers dictates one's thoughts in a way that is vaguely reminiscent of the telegraph in an earlier era.

The screen is a portal. In order to understand that portal, one must discern the ways in which the screen alters the figure-ground gestalt. In the past, the screen in the form of television has been compared to print and found lacking. It is correct that print and the screen foreground different elements. Setting aside the comparison of print and screen, the screen as we now know it takes many forms. The screen integrates images, sound, and movement in addition to text, in a way that is both comparable to and yet different from ordinary experience. More than any other communication technology, the screen incorporates the full range of senses with the exception of smell and taste.

It is not just the isolated eye or the ear that is called into play. Nor can the experiences of the screen be reduced either to dialogue or to imagery. The screen presents a synesthetic experience capable of playing with and exploring the limits of the human sensorium. McLuhan attempts to capture that quality of synesthesia by his use of the terms *mosaic* and *immersive* to describe early television. The experience of the screen integrates the ear and the eye in a way that is related to one's unmediated experience. Yet in ordinary life the eye and the ear are never joined that tightly, the link is never that explicit. It is the comparison of ordinary experience and the mediation of that experience that will reveal the portal of the screen.

Though it appears at first glance that the screen, like the book, creates a new awareness of space, the screen does not foreground space. As Dorfles explains, the screen draws on both time and space to create a verisimilitude that is unlike anything one might experience in ordinary life. In the world without mediation, one's orientation is defined by a particular perspective, is constrained by the physical location of the body. It is not possible to omnisciently manipulate time and space. One's lived experience does not provide the rapid shift of multiple vantage points.

Nor does the screen foreground dialogue, though it certainly offers opportunities to explore the dynamics of conversation, to replay in a way that is simply impossible in ordinary life. What fundamental element of the screen surfaces as the defining feature? What makes this experience different from other forms of mediation, regardless of the size, definition, or type of screen? The screen integrates rhythm, space, and sound to create a felt sense of motion, an empathic element that conveys information not simply through words nor through imagery but through an altered duration.

The true nature of that portal will never be realized if the images of the screen are considered to be just as real or unreal as any other perceptual experience. Perhaps it is merely whimsical to suggest that all lived experience can be reduced to mere semblance, to presume the ontology fails, to argue that one appearance is as real or as unreal as any other. Postmodern critics posit the collapse of the relation between sign and referent, arguing that "there is little difference between the 'world' that [media] has constructed and the material world where [media] is ubiquitous" (Hay, 1992, p. 357). Based on this view, the mediated portrayal of the world, the cataclysmic images of storms or war, the displays of international politics, are no more or less real than the immediacy of one's lived experience. But if there were no difference between mediated and non-mediated events, between the media construction of a world and one's lived experience of that world, the screen could not serve as a portal to deliver one to a different place. Simply put, there would be no difference. A collapse of the relation between sign and referent would render such a distinction unintelligible.

The mediated newscast is quite different, in crucial ways, from one's experience of lived events. The manipulation of images through computer graphics, camera angles, editing, and digital display creates a scene that is both strikingly real and yet nothing like what one experiences without the benefit of such mediation. Time, space, motion, and sound are altered to create a composite that is distinctly different from ordinary experience. If examined carefully, the contrast between lived experience and the technological mediation of that experience will reveal the ways in which figure and ground are altered.

It is one's experience of existing in an ordinary time and space that provides the intriguing benchmark against which to compare and analyze the electronic portrayal of a world both familiar and strange. Postmodern commentary acknowledges the strangeness of that media portrayal, pointing out the technological aura of alienation and fascination. The mistake is to assume that either the visual display of the screen or the terms of the printed text constitutes alienation. The audience does not become a voyeur or merely a consumer, nor is the world simply reduced to spectacle. Reality does not become the random play of surfaces, though something far stranger may be going on.

The basic features of the world remain unchanged. There is time and space, a horizon, near and far, foreground and background. Self and other exist in some relationship. Mediation, in any form, plays with these basic elements. Print provides new ways to structure information. The telephone changes one's experience of conversation. The screen alters our shared experience of time and space. Postmodernism may argue that media blur distinctions between real and unreal, self and other, but a more thorough explanation is needed to explain the consequences of such a blurring.

Understanding the way in which different communication technologies alter the figure-ground can help one recognize that blur for what it is: the reconfiguration of a perceptual world that appears intimately familiar yet somehow foreign. The horizon still exists, but one's understanding of that horizon is altered. Alienation may be the result, but one can also be fascinated, enticed, or engaged. All media have the potential to create bonds between people as well as to alienate people. Alienation tells only part of the story. The book can be experienced as an intimate form of isolation. The telephone can be understood as a way to reach out and touch someone but only at a distance. The portrayal of world events on the evening news paradoxically positions the world within our grasp in a way that exceeds our grasp. It is simplistic to presume that, as a consequence of media, one merely sees the world differently. More than the eye is called into play. The screen reveals a very different experience of the world that melds sight, sound, and movement.

The Gestalt and the Communication Matrix

Print altered the reader's awareness of inner experience; the screen alters one's awareness of outer experience, of the self in the social world. It is not just that the mediated portrayal of the world differs in striking ways from one's lived experience; if that were all there was to it, we could enjoy different aspects of the technology and let it go at that. But when a shifting figure-ground gestalt alters the communication matrix, new forms of awareness are created and consequences follow. Consider how the world appears on the screen: the sound, rhythm, tempo, and movement. When one's use of media becomes ubiquitous, that montage of image, movement, and sound is internalized. The residue of phrasing and tempo shapes one's subliminal response to the world. If you doubt that the impact of a particular technology could be that profound, consider the impact of writing. Script and print changed how records were kept, and also influenced the nature of community and identity. The ability to record one's thoughts made it possible to reflect on the life of the mind, to fix one's thoughts on the smooth page and compare those thoughts to past and future dreams. Writing and printing revolutionized communication. Electronic media constitute yet another communication revolution.

The book foregrounds an understanding of space based on typography; that awareness of font, page layout, and chapters provides an opportunity to explore not only the structure of words on the page but the structure of one's thoughts. The ability to fix the words on the page meant that the mind was freed from the burdensome task of remembering, freed to invent new ways of saying things, to construct a visible space for one's ideas. The author carefully considers the organization of sections and chapters; the reader notes that the organization of the book provides clues to the writer's intentions. It is inherent in the technology of print that this organization will be spatial and necessarily visual. One may choose to use a particular technology, but one cannot choose how that technology will privilege communication.

Each medium emphasizes the figure-ground in a distinct way. One may use a technology without being aware of how a particular medium foregrounds certain elements and minimizes other elements. That is to say, one may use the technology without explicit awareness of its impact. As noted earlier, song cannot foreground space in the way that print does. This obvious point appears to be inconsequential until the implications are made explicit. If one uses a technology often enough, a technology that foregrounds certain elements over others, one cannot help but experience the world through the bias of that medium. The technology creates a residue, a filter, a way of perceiving. McLuhan argues that media are an extension of the human nervous system. Whether he intended his point to be taken literally or figuratively, it is clear that technology enlarges the range of vision, opens the horizon for distant conversations, extends the reach, alters tempo and tone, and, in the process, inevitably alters one's perception as well.

When one's attention is focused on the message, the e-mail, the advertisement, the printed page, or the telephone conversation, it is difficult to notice this pervasive quality. It is an ironic point that the ability of each technology to alter the figure-ground gestalt is, in turn, minimized by the message. One remembers the conversation on the telephone, not that certain elements of the conversation were missing. One uses electronic mail to communicate instantaneously over great distances without considering how e-mail is distinctly different from both the telephone conversation and the handwritten letter.

Consider the biases of print, music, or video. The more thoroughly one is schooled in a particular medium, the more difficult it is to grasp the ways in which that technology influences one's perception. The more deeply one is ensconced in a particular medium, the easier it is to experience the world through that figure-ground. The more frequently one experiences a particular medium, the more pervasive the influence of that bias. Perhaps it was easier to detail the transition from orality to print because there were a limited number of technologies in use at that time. In order to grasp the bias of a particular medium, one needs a different vantage point, a new perspective that calls the bias of the old order into question. Television provided such a perspective for a print-dominant culture. The computer could provide such a perspective for television but for the fact that soon television and computer will merge into one technology that is not broadcast, not video, not print, not quite television or computer in their current form. If it were just print, or radio, or telephone, or script, or just television, it might be easier to clarify the ways in which perception is reconfigured.

McLuhan was a man of letters. As a professor of English, his bias was print. No doubt he was aware of that, just as he was aware that the second half of the twentieth century marked a cultural breakpoint. He could discern the transformation in its earliest form. There are at least two possible reasons why this man, schooled in the medium of print, could escape the print bias to discern the emergence of a new form. His study of English literature included Chaucer, and his daily practice included the Catholic mass. Both Chaucer and the Catholic mass are based on oral rather than literate forms of expression. Both his intellectual pursuits and his faith provided an opportunity to experience the power of orality. Those forms might have provided a counterpoint for this man of letters. Because our experience of the world today is highly mediated, it is difficult to find that counterpoint. In a typical day one deals not with a single communication technology but with many. Now speech is mediated in diverse ways, and writing can mean anything from text to hypertext. In addition to speech and writing, there is the incessant interplay of visual forms.

In this mix of speech, text, and image, there is a point of reference that guides one's awareness. The telephone is not one's primary means of experiencing the voice of the other. The images of the screen are not the first way one gazes at the world. The words of the text do not provide an introduction to words. It is the lived body that mediates one's experience of the world; the human sensorium discovers and rediscovers one's relationship to that world through the interrelationship of sight, sound, touch, and movement.

Perhaps in the future there will be humans who are entirely dependent on a technological interface, people who never hear the human voice, never stand in some relationship to a physical horizon, never hear the sound of the wind in their own ears, or see the form and color of a summer field. It is difficult to know whether such a person might be considered fully human. There is the bodily way in which one is situated as a subject, the sheer physicality of the hand that reaches to caress and is, in turn, caressed by a world that is both object and subject. To forget that uncontested immediacy, to forget the way that the gaze and the hand make the world their own, is to lose sight of one's human legacy. It is this bodily way of being in a world that provides the necessary counterpoint. This is Merleau-Ponty's point.

The contrast between lived experience and the mediation of that experience provides the necessary counterpoint needed to understand how mediation alters one's experience. We are aware of the contrast between what might be referred to as "ordinary" or lived experience and the technological mediation of that experience. Why else would one prefer technology that creates richer sound, brighter colors, more lifelike images? Without the phenomenology of the lived body, there would be no making sense of the images of the screen, the sounds of the voice on the CD, the letters on the smooth page. The technological mediation of the world was created to replicate, and in some cases to intensify, human experience.

Few people today would be satisfied with the small, snowy, black and white screen of early television. The iteration of the screen has led inevitably to brighter, sharper images and sound. The pleasure comes from creating images that are beautifully real on a screen that is larger than life with special effects and sound effects that heighten the intensity. The physical space of the book could not evolve in the ways that the screen has evolved. Over centuries the format of the book has not changed much, whereas the screen has reinvented itself in a matter of decades. In another century, the screen will not resemble what is available today.

The book could never replicate the physical world in the way that the screen does. Because print involves fewer of the senses, its impact on the human sensorium is quite different from the screen. In important ways, the book and the screen are antithetical. As various scholars have pointed out, print is a visual medium. Careful consideration of the screen reveals that it does not work effectively as a visual medium. The screen links auditory and visual elements with motion.

That kinesthetic portrayal of the world is unprecedented. No technology prior to the screen was able to capture the world in motion in this way. To live in a world dominated by the screen is vastly different from living in a world dominated by books. The shift in the figure-ground alters the communication matrix by privileging certain forms of experience over others. The way in which the screen privileges motion has an impact on one's experience of time and space.

The book created a way to organize and record one's thoughts within a contained space. The reader expects the words of the book to be arranged in a logical structure. Volumes have been written to address what is meant by logical structure. The screen usurps that logical structure by portraying thoughts and feelings in a very different form. This shift in mediums contributes to a corresponding shift in awareness. That shift in awareness can be understood by investigating the communication matrix.

Traditionally, the senses are discussed as if they are separate entities. Reference is made to the eye versus the ear as if these forms exist independently of each other. Primary orality relied on both the eye and the ear; print internalizes the auditory component. The reader imagines the voice of the author speaking in her head or plays out the different voices of the characters of the narrative. In actuality, the senses may be deeply interrelated. It is time to move beyond "simple models of cerebral organization" that view the senses as separate and distinct. In an article entitled "Joined at the Senses," Bruce Bower quotes Robert J. Zatorre's argument that "interactions across sensory receiving areas in the cortex may be more widespread than previously suspected" (2001, p. 204). Can the senses be considered as separate and distinct? Some neurophysiologists object:

> It's time to drop the assumption that people and other animals perceive the world through separate sensory channels of vision, hearing, touch, taste, and smell....Perception isn't grounded in simply seeing what is visible or hearing what is audible, in this scenario. Instead a person simultaneously exploits different sources of energy, such as light radiation and acoustic pressure waves, as he or she strives to attain certain goals, such as understanding speech. (p. 204)

The term *simultaneously* emphasizes the ways in which the auditory, visual, and tactile senses contribute to the perceptual gestalt. It is possible that the senses may be both intertwined and distinct in ways not yet fully understood. The concept of multisensory perception is relevant because it has been argued that communication technologies can alter the ratio of the senses by privileging certain types of information over other types of information. I am not suggesting that communication technologies alter the physiology of perceptual processes. Rather, perceptual awareness can be influenced through mediation. A better understanding of the physiology of those "interactions across sensory receiving areas" might challenge existing notions of how mediation influences perceptual awareness.

The photographer sees the world differently than one who has never looked carefully through a lens. The dancer's sense of the body's form is different from what it is for someone not trained in dance. The book shapes perception as well, as does the screen. That influence can be understood, to some extent, by recognizing the ways in which different technologies privilege certain elements of perception over others, by understanding how a particular medium shifts the figure-ground.

It is worth noting that the concept of multisensory perception arises during that time when the dominant medium of the screen is offering a multisensory mix. Unlike print, photography, telegraph, or telephone, the screen mixes different sensory forms to create a kinesthetic sense of motion. It is that multisensory mix that contributes to the unique powers of the screen. Turn off the sound and watch the screen to discern the importance of the auditory component. Turn your back to the screen and listen to discern the importance of the visual component.

The kinesthetic element arises from the joining of the auditory and visual component. This medium mixes different sensory channels to link space, sound, and motion. The result is an awareness of time and space based on rhythms that are not necessarily part of one's ordinary experience. Music has always offered a way to experience and experiment with different tempos. The screen links tempo to a visual panorama that is distinct from both music and from one's lived awareness of being in the world. Inadvertently, a new space conception has been created. We live within that space conception now; we experience but do not understand it.

Differing Space Conceptions

I borrow the term *space conception* from the work of Sigfried Giedion. In an article titled "Space Conception in Prehistoric Art," Giedion identifies the way in which humans give psychic form and expression to the physical space they inhabit: "The effect of this transfiguration, which lifts space into the realm of the emotions, is termed space conception. This space conception portrays man's relations with his environment" (1960, p. 73). Contrast the way that Euclidean geometry represents space with the concept of space characterized by Einstein's relativity. Compare the space conception of artists such as Klee or Kandinsky with the space depicted by the realism of earlier centuries. These comparisons capture very different ways of thinking about and portraying space. In *Space Perception and the Philosophy of Science* (1983), Patrick Heelan argues that "it is not at all evident that the space we perceive is everywhere at all times Euclidean" (p. 258). One's experience of space is based on tacit, shared cultural assumptions. According to Heelan, "the environment we experience tends to become the surroundings of those technologies that interface between us and our World" (p. 248). Giedion maintains that such a conception of space "develops instinctively, usually remaining unknown to its authors. It is just because of its unconscious and, so to speak, compulsive manifestation that a space conception provides such an insight into the attitude of a period to the cosmos, to man, and to eternal values" (p. 73). Giedion argues that such attitudes toward space "change continuously, sometimes by small degrees, sometimes basically" (p. 74). The way one sees the world involves much more than the physiology of the eye. Vision does not merely discern the world as it is, like some separate object fully defined in advance. To view an object on the horizon is to experience one's relationship to that object.

Much of McLuhan's work can be understood as an exploration of differing space conceptions. He contrasts the perspectival visual space, associated with the emergence of the alphabet, with an earlier, preliterate acoustic space. That perspectival visual space can be referred to as *pictorial vision*. According to Heelan, "pictorial vision consists in the art of seeing objects as pictorial, that is, as constructed according to rational (i.e., in this case, geometrical) principles out of basic visual pictorial elements, such as points, segments of lines, patches of plane surfaces, and the color and light on these patches" (1983, p. 42). Reality, as Heelan explains, is not exclusively pictorial. Neither the "facts of perception" nor "philosophical reflection" support the reality of pictorial space. Science and phenomenology offer a more thorough way of conceptualizing space.

Nor is it wise to assume that all peoples and cultures share our understanding of pictorial space. Heelan argues that "whole communities have been accustomed...to see their World, or horizons of their World, as spatially organized in ways that we find hard to describe at the conscious level without using the language of illusion and distortion" (p. 27). Giedion argues that space conceptions change slowly over time. How one experiences space today may differ in important respects from how a person experienced space in an earlier century. Though few might be able to appreciate non-Euclidean geometry, many can appreciate the vision of those artists who intuitively understand different space conceptions. Compare the pictorial realism of the eighteenth century with the depiction of space portrayed in twentieth-century cubism.

McLuhan referred to pictorial space as a bounded and rational system that offered a favored point of view. All objects are situated in a rational relationship to that singular viewpoint. As McLuhan explains, objects "must be leveled off by some continuous narrative, or be 'contained' in some uniform pictorial space" (1994, p. 289). According to Carpenter,

> Literacy creates a "middle distance"; it separates observer from observed, actor from action; it leads to single perspective, fixed observation, singleness of tone, and introduces into poetry and music the counterpart of three-dimensional perspective in art, all, of course, artistic expressions of the Western notion of individualism, every element being now related to the unique point of view of the individual at a given moment. (1966, p. 214)

McLuhan argued that this rational concept of space was encouraged first by the book and then by film, to the extent that film borrowed the narratives of print. There must have been other influences as well. Physicists, poets, and mapmakers can challenge a culture's space conception by introducing new ways to represent space. It was McLuhan's point that television did not exhibit a pictorial space. How he might have discerned that, with television still in its infancy, I cannot know.

In another culture with different influences, pictorial space would not emerge as a dominant space conception. Consider traditional Chinese landscape painting that designates not one favored point of view but a range of ascending vantage points depicting shifting perspectives. The literacy of Eastern cultures, arising from iconographic rather than phonetic forms, contributed to a vastly different space conception as well as very different ways of conceptualizing the self in the social world. The rationality of pictorial space implies a very different kind of observer than the observer of the traditional Chinese landscape. Depictions of space are always twofold, portraying what is to be viewed as well as the implied location of the viewer. Though McLuhan was concerned with the space of the eye and the ear, with visual and acoustic space, he also discerned the relationship between a culture's space conception and a culture's view of the self. The ways in which a culture conceptualizes space implies something about the ways in which that culture defines the self.

Presume for a moment that McLuhan had it right, that the book contributed to a rational, pictorial space, which situated both observer and observed within a system of logical relationships. Compare that rational space to the space of the screen, which does not adhere to the pictorial. The camera moves; the scene swims in and out of focus. The zoom lens references a space where scale and distance can be instantly altered. Sunlight and shadow are used to depict space as well as to capture the passage of time. The relationship of near and far is neither fixed nor constant. The screen can utilize many different kinds of images. There is the display of typography offered by satellite communication and the inner workings of the aortic valve displayed on the screen for the surgeon. There are the strands of DNA that can be set in motion with a twist of the head, as displayed through virtual reality, and the DVD portrayal of a world too beautiful to grasp.

Pictorial space is usurped by the screen, which can sustain, stretch, and alter one's awareness of space. The screen can display numerical experiments in mathematics such as the Mandelbrot technique, used to locate complex numbers that do not increase to infinity (Friedhoff & Benzon, 1989, p. 165). There are computerized object-based systems utilizing geometry and three-dimensional graphics that combine elemental shapes (p. 90). Space, time, and distance are applicable not only to pictorial space, but also to the viewer. Contrast the observer implied in that pictorial space with an observer of the screen. The logical relationships of pictorial space are gone. There is no necessary horizontal or vertical axis. Instead of a constant scale, the eye seemingly turns in different directions simultaneously. There is the athlete leaping for the net, the cumulus clouds racing across the horizon, the mathematical structures displaying relationships never before envisioned. In constant motion, the observer understands that the scene can transpose near and far in the blink of an eye.

It is easy to assume that the term *space* as used here refers to visual space. But space also exhibits auditory and kinesthetic properties. That our culture thinks of space as visual rather than auditory merely indicates one dominant feature of our space conception. By contrast, the space of the screen is visual, auditory, and kinesthetic. The term *acoustic space* indicates a way of experiencing space that is based on the ear rather than the eye. The blind would have little difficulty understanding the concept of acoustic space. For a visually dominant culture it is more challenging.

Carpenter presents an account of acoustic space in an article titled "Image Making in Arctic Art" (1966), which details the language, art, and customs of the northern Canadian Inuit. Carpenter emphasizes the way in which survival on the formless frozen arctic tundra requires a distinctly different awareness of space, based on the sounds of snow, ice, and wind. For the Inuit,

> Auditory space has no favored focus. It is a sphere without fixed boundaries, space made by the thing itself, not space containing the thing. It is not pictorial space, boxed in, but dynamic, always in flux, creating its own dimensions, moment by moment. It has no fixed boundaries; it is indifferent to background. The eye focuses, pinpoints, abstracts, locating each object in physical space, against a background; the ear favors sound from any direction. (p. 220)

His article provides a striking counterpoint to western depictions of space. The western orientation, based on a vertical and horizontal axis, is disavowed in favor of multiple perspectives. There is no favored point of view. Carpenter argues that the "value we place on verticality...stems from the strength of literacy in our lives" (p. 219). When space is no longer conceptualized as pictorial, foreground and background are not fixed, and multiple perspectives are possible. Space structured in a non-optical way conforms to very different rules. The logical relationships of objects within the frame, the orientation of up and down, the favored point of view no longer hold.

This view of acoustic space suggests a concept of self quite different from the implied observer of pictorial space. Carpenter's essay on the Inuit establishes a connection between a specific space conception and a view of the self that is quite distinct from Western views of the individual. In discussing acoustic space, Carpenter describes a view of the self without inner dialogue, without reference to proper pronouns, without the accruements of individuality as understood in our culture. For the Inuit, "thinking and speaking are one: there is no purely inner experience" (p. 212). The solitary self is not privileged; their experience is collectivistic. Identity is determined by one's relationship to others. Just as auditory space provides no favored point of focus, their carvings display multiple perspectives. There is no one correct way to view a carving, just as there is no one correct way to view the self. The identity of each person is neither singular nor fixed. A member of the tribe may have been known by several different names.

> They do not reduce the self to a sharply delimited, consistent, controlling 'I.' They postulate no personality 'structure,' but accept the clotted nature of experience—the simultaneity of good and evil, of joy and despair, multiple models within the one, contraries inextricably commingled. (p. 224)

This description of collectivistic identity is quite similar to Ong's and Havelock's description of primary orality. In a highly literate culture, this way of being is reserved for young children, creative geniuses, and lunatics.

Whereas we might speak of weak ego boundaries to describe the person who displays distinctly different personalities, the Inuit, by contrast, would find the use of the proper noun "I" quite confusing. It is reassuring to establish a sharp contrast between one's culture and the other and to then romanticize the other as completely unavailable to one's own experience. In fact, the collectivistic experiences of the Inuit are evident in Western culture just as the Inuit, in turn, displays an awareness of the autonomous self. It is a matter of what is emphasized in each culture, rather than incommensurable experiences. These distinctly different views of the self characterize two points on a continuum.

As Levinson explains in *Digital McLuhan* (1999), auditory space is "a world of no boundaries in which information emerges not from fixed positions but anywhere and everywhere" (p. 45). Levinson makes the point that "acoustic space is now most found in the online alphabetic milieu of cyberspace" (p. 46). It is correct that the layout of the book and the layout of cyberspace are profoundly different. Cyberspace is of the screen rather than of the book. Acoustic space is merely one of many different ways of conceptualizing one's relationship to the external world. It would be a mistake to presume that there is pictorial space, and then there is acoustic space. The acoustic awareness of the Inuit perhaps differs from the spatial implications of cyberspace. Here again, the ghosts of orality and secondary orality are invoked in an attempt to discern the outline of a new era.

The great artists of the twentieth century have introduced diverse ways of experiencing space. Consider the sculptures of Giacometti and Moore, the paintings of Picasso, Braque, the dreamlike landscapes of Miró, the work of the abstract expressionists, to name but a few. A detailed analysis of the transition from orality to literacy confirms that print had a profound impact on collective awareness of space. So too might high-rise buildings, superhighways, long-distance telecommunications, and electric lights. It is understandable that, on the frozen tundra, the Inuit would conceptualize space as auditory. Their lives depended on the sound of the wind and the audible crack of the ice beneath their feet. To reduce cyberspace to auditory space is to lose sight of the complexity implied by Giedion's use of the term *space conception*:

> Man takes cognizance of the emptiness that girdles him and gives it a psychic form and expression. The effect of this transfiguration, which lifts space

> into the realm of the emotions, is termed space conception. This space conception portrays man's relations with his environment. (1960, p. 73)

The singular concept of auditory space does not adequately capture what is emerging with the culture of the screen.

The dialogue has been structured around a series of dichotomies. On one side there is literacy, the eye, and pictorial space; on the other, primary orality, the ear, and acoustic space. As with all dichotomies, these sets of terms both ordain and constrain the dialogue. It was McLuhan's weakness to draw heavily on this dichotomy. It was his strength to understand the extent to which this dichotomy could foreshadow a new era. Levinson and others are quite correct that cyberspace is profoundly different from the pictorial space print implies. But the dichotomy misleads to the extent that it suggests a simple return. It is not possible to go back. Orality, even secondary orality, is no longer an option, if it ever was. The future rushes toward us. Insofar as the dichotomy designates two choices and implies the possibility of a return, it misleads.

Secondary orality was a useful concept earlier in the twentieth century for helping one think carefully about the implications of mass media. It is now as outdated a concept as the term *mass media*. The bifurcation of eye and ear has guided the dialogue for too long. The senses are intertwined. Space is both acoustic and visual. Our current highly mediated environment reveals a space conception that is both sequential and nonsequential, pictorial in some ways as well as cubist, relative, chaotic, non-Euclidean. There are more geometries than Euclidean and more logics than the Logical Positivists ever imagined. The screen does not work according to the dichotomy of eye and ear, of pictorial and acoustic space. If the dialogue of orality and literacy has become constrained, it was nevertheless necessary. It is now possible to transcend the very dichotomy that guided the discussion.

A World in Motion

Motion distinguishes the space of the screen from the book. With the book, time and space are of necessity fixed on the smooth page. The imagination of the reader calls forth the details contained in the printed words. With the screen images move, whether those images be words, landscape, or human forms; nothing is static. Whereas the book emphasizes the visual sense, the screen integrates sight, sound, and motion. The distinction of eye and ear that served to explain the transition from orality to literacy cannot address the altered duration of the screen. If we are to fully understand the nature of the screen that has become such a pervasive part of our daily lives, we must not separate image, sound, movement, and word. Rather, we must focus on the way in which visual and auditory forms contribute to a felt sense of motion that is different from anything prior to the screen. To explain the unique features of the screen, I will draw on the work of two pivotal figures from the mid-twentieth century, both art historians who understand the implications of new forms of expression.

First, the work of Erwin Panofsky details what is unique to the medium of film by contrasting the screen and theater. The fundamental element of motion that distinguishes book and screen was evident with the earliest silent films. That element of motion was a fundamental feature of all screens. Panofsky's discussion of film and theater prepares the way for a more thorough understanding of the screen.

Second, the work of Dorfles establishes the ways in which that felt sense of motion, first identified with the screen, is now a pervasive part of our everyday lives. New forms of communication arise as part of a larger social environment. The impact of a new medium is expressed through the interaction of technology and social environment. As we shall see, the impact of the screen mirrors other technological forms that alter one's experience of tempo, rhythm, and pace. Though Panofsky's and Dorfles's articles were first published in 1993 and 1965, respectively, their ideas provide an intriguing glimpse of the future.

In an article titled "Style and Medium in Motion Pictures" (1995), Panofsky focuses on the essential difference between moving pictures and theater. Perhaps it was his training in the arts that enables Panofsky to grasp the uniqueness of film. He is first and foremost interested in the fact that film is capable of creating a sense of movement. He argues that, even with the advent of the talkies, film remains a picture that moves: "Contrary to naïve expectation, the invention of the sound track in 1928 has been unable to change the basic fact that a moving picture, even when it has learned to talk, remains a picture that moves and does not convert itself into a piece of writing that is enacted" (p. 100). He seeks to define the essential features of each medium, just as a medium theorist would. A careful consideration of the stage reveals that the dramatic performance is limited by its sheer physicality. The scene is bound by the constraints of that stage and is capable of moving forward or backward in time only through the words and gestures of the actors or through carefully orchestrated scene changes. Though the live dramatic performance offers many aesthetic charms, the rapid manipulation of space is not one of them.

Understanding what the stage foregrounds helps to reveal the unique features of the screen. Panofsky argues that "the theater has the advantage that time, the medium of emotion and thought conveyable by speech, is free and independent of anything that may happen in visible space" (p. 96). The stage foregrounds the physical presence of the actors, embodied through voice, gesture, and posture. The screen foregrounds movement. "With the movies the situation is reversed....[The spectator] is in permanent motion as his eye identifies itself with the lens of the camera, which permanently shifts in distance and direction" (p. 96). Theater and film are fundamentally different because of the camera's ability to alter the viewer's perception of space as being

> ...as movable as the spectator is, as movable is, for the same reason, the space presented to him. Not only bodies move in space, but space itself does, approaching, receding, turning, dissolving and recrystallizing as it appears through the controlled locomotion and focusing of the camera and through the cutting and editing of the various hosts—not to mention such special effects as visions, transformations, disappearances, slow-motion and fast-motion shots, reversals, and trick films. (pp. 96–98)

In the theater, space does not easily recede, dissolve, or recrystallize. With the camera, all of this and more is possible.

The screen today does not depend solely on the camera. From the beginning to the end of the twentieth century, both camera and screen have advanced to include computer-enhanced digital and graphic imaging that is nothing like the earliest talkies. Despite these significant technological innovations, the varied iterations of the screen continue to display what Panofsky refers to as "*dynamization of space* and, accordingly, *spatialization of time*" (p. 96). These terms characterize the way in which space is set in motion and time is revealed through the manipulation of space. It is correct that space moves on the screen, but bodies also move in space. The *dynamization of space* and *spatialization of time* capture and transport the viewer, creating a sense of motion that is kinesthetic, visceral, felt. The term *motion* is related to the term *emotion*: "the movies have the power, entirely denied to the theater, to convey psychological experiences by directly projecting their content to the screen, substituting, as it were, the eye of the beholder for the consciousness of the character" (p. 98). It is the emotional quality of that kinesthetic motion that distinguishes the screen.

Some images are less laden with emotion than others. The computer-aided design sequence will not move one in the way that images of starving children arouse compassion. Just as there are different types of books with different purposes, there are different types of screens. Nevertheless, to experience the computer-aided design sequence is to position and rotate an object in space, to display internal and external elements from varied perspectives. The viewer must shift with those varied perspectives, must establish his orientation and maintain a frame of reference. This kinesthetic activity draws on cognitive and emotional processes in ways, to degrees, that are vastly different from any medium prior to the screen. Panofsky understood this well in advance of computer-enhanced design. As McLuhan suggests, it is the artist, or at the very least it is training in the arts, that helps one to discern the terrain of new media.

Panofsky's *principle of coexpressibility* challenges the distinction of eye and ear. For Panofsky, the acoustical element is not detachable from the visual. He theorizes that film operates according to the principle of coexpressibility, insisting that:

> In a film, that which we hear remains, for good or worse, inextricably fused with that which we see; the sound, articulate or not, cannot express any more than is expressed, at the same time, by visible movement. (p. 100)

Note the reference to "visible movement." Panofsky's principle of coexpressibility suggests that the fusion of visual and auditory elements is expressed through motion. The screen is more than just a visual medium. Nor can it fairly be identified as an acoustic medium as has been suggested by the term *acoustic space*. The screen is unique in the way that it foregrounds not the eye or the ear but a new kind of movement which links eye and ear to a degree that is strikingly different from ordinary experience. As J. Mallory Wober explains, "when...one sees portrayals of others and their posture and movement and one has some sense of immediacy of or involvement with the portrayed reality, one's own system of balance and position is called into play" (1988, p. 166). It is not just the case that the camera supports what is provided through sound. Rather, new complexities emerge as sound and vision are fused to create a sense of movement.

Without the benefit of technological mediation, one perceives the world from a fixed perspective. There is no omniscient vantage point from which to observe shifting or multiple scenes. One cannot jump instantly from one scene to another. In ordinary experience, one does not orchestrate and script images to sound. One's most profound experiences can be disrupted by the ringing of the telephone. Heartfelt memorable moments rarely come with appropriate musical accompaniment. The screen can do all of these things. One consequence is that it is possible to perceive, feel, and think about the world in ways that were never before possible. Panofsky's principle of coexpressibility provides a conceptual way to address this crucial integration of space, sound, and motion. The coexpressibility of visual and auditory elements intensifies one's awareness of movement.

Panofsky was concerned with film. I am concerned with the screen, which includes film, video and DVD, cyberspace, video games, virtual reality, graphic imaging and design. Not all of these screens utilize sound, though auditory elements are increasingly incorporated to unify the experience of the screen. Nor do all screens display Panofsky's principle of coexpressibility, though it could be stated that, as a general rule, meaningful commentary must integrate visual and auditory components. Panofsky was concerned with the aesthetic qualities of film, which was the first medium to capture the visible world in motion. With advancing iterations of the screen, it is now possible to capture not only the art of fictional narration but instantaneous commentary from across the globe, creating a technological space-time that is both live and profoundly artificial.

It is possible to overlook the movement displayed on the screen, the sequencing of cuts, the shift from one scene to another. Observing an interview, debate, or discussion, it is difficult to explain exactly how that discussion is being contextualized within an artificial space-time. That movement in space and time constitutes a significant part of the message, which can be influenced by factors as subtle as the gesture of a hand or as obvious as a close-up emphasizing the speaker's face. The screen can reveal those aspects that print omits: the flow of image, sound, and movement. Ideas communicated via the screen are contained in neither the image nor the word but emerge through a fusing of the two. The result is a syntax of motion.

Following the tradition instituted with the earliest talkies, the screen unifies the senses with a hypnotic power that subtly alters our experience of space, time, and duration. Though that syntax can never be fully captured, it nevertheless serves our purposes to investigate the logic of perceptual awareness as mediated by the screen. What are the consequences of the camera's movement? What is the significance of gesture? How is time altered and one's experience of duration transformed? Such an analysis will reveal the unique features of the screen.

The motion of the screen is but one of many ways in which technology has created an altered duration that is largely taken for granted. In an article titled "The Role of Motion in Our Visual Habits and Artistic Creation" (1965), Dorfles explains how a new experience of mechanized motion "must be considered one of the most significant dimensions to which present-day existence is subject" (p. 41). Dorfles understood that the twentieth century had created a "life in motion," defined in large part by a dynamism that was at odds with natural biological rhythms. He drew on examples of automobiles, jets, bobsleds, motion pictures, and escalators, among others. Dorfles explores the mechanical rhythms of the modern era:

> While man has been conditioned ever since his appearance in the world, and is constituted in all his physiological structure to obey and respond to a "cosmic rhythm" (bound obviously to breathing, to cardiac pulsation, to the mysterious rhythms of the universe, alternating like day and night, the tides, the months, etc.), it is probable that only in our era has he found himself in

> contact with mechanical rhythms which interfere profoundly with his interior rhythms. (p. 45)

Dorfles was interested in the pace of modern life as evidenced in speed, tempo, and dynamism. He realized, as did McLuhan, that one may be scarcely aware of such changes and, nevertheless, be profoundly affected.

One rarely thinks, when taking off on a plane, that the experience of speed, the thrust of the engines, the landscape rushing by beyond the small window all have an impact on one's physiology. The rush of the expressway may create a sense of excitement or frustration, but rarely does one stop to reflect on the ways in which that experience is physiologically different from earlier experiences of walking or riding in a slow-moving carriage. The number of years humans have spent speeding along the expressway is miniscule compared to the number of years humans have traveled at the pace of a slow walk. The experience of the plane and expressway are just part of one's ordinary life. Yet the rhythms of such technologies interact with eye, breath, heartbeat, with one's kinesthetic awareness of being in the world.

Though Dorfles does not use McLuhan's term *sense ratios*, he does emphasize the altering of the sensorium. He argues that "certain events...set up important transformations in [one's] sensory, perceptive and proprioceptive activities, causing a reaction throughout entire areas of his productive and communicative activity" (1965, p. 41). He understands that motion "is capable of favoring and intensifying the empathetic element...because, through the transmission of a peculiar rhythm, it 'sets in motion' the innermost rhythmic structures of the human constitution" (p. 45). If motion can favor and intensify the empathetic element, then what of the motion displayed on the screen? Though few people can experience first hand the pace of the speed skater, many can resonate to that experience broadcast live on a wide screen or high-definition screen.

Dorfles suggests that the creation of a new dynamism alters how one experiences, sees, and represents the particulars of their experience, despite the fact that one's response to that altered duration may occur below the level of reflective awareness. In turn, I argue that altered duration impacts not only one's "sensory, perceptive and proprioceptive activities" but one's relationship to others. Technological interventions, which alter how one experiences, sees, and represents the particulars of experience, also alter one's relationship to others in the social world. The challenge is detailing the connections between that altered motion, the shift in "sensory, perceptive and proprioceptive activities," and corresponding changes in the social domain. This chapter has investigated the important ways in which the screen creates an altered duration. Subsequent chapters will establish the ways in which that altered duration impacts the social world.

We notice only that life seems to move more quickly, that there is less time. We are fascinated by time, hurried and rushed, stretched in ways we find difficult to explain. It seems we move to a different tempo though the implications of that point escape our grasp. One's orientation to others, the glance, gesture, and measured response, are examples of communicative acts that are defined in subtle ways by distinctly human rhythms. Much of our experience now includes an artificial sense of dynamism that draws on rhythms of a very different sort.

The screen enhances and projects an altered duration that extends one's reach and challenges one's grasp of what it means to exist in balance with others. Dorfles argues that such chronological and spatial coercions "are not considered, as is usually true in the other arts, as 'imaginings,' as 'fantasies,' but often, because of their apparent realism, as if they were the mirror of an effective reality" (p. 47). Dorfles understands that the images of cinema and television are powerful because those images can be set in motion. This essential feature of motion contributes to the verisimilitude of images that seem not only to move but also to breathe and live. Captivated by that sense of verisimilitude, he writes:

> The information which the spectator receives through the film accustoms him to jumps in time, to retrogradations, to chronological and spatial coercions, so that at the end of the performance he finds himself in possession of an accumulation of data entirely different from that which the experience of normal life furnishes him. (p. 47)

The screen captures and enhances nuances that, under other circumstances, might appear irrelevant. Those who argue that there is no reality beyond the mediated event misunderstand the power of the screen. Only by comparing the screen to the lived world can we grasp the impact of such dynamism. Panofsky explains the unique features of the screen. Dorfles extends that discussion to include the broader technological environment. Each scholar complements the other. Their respective views confirm the ways in which the "dynamization of space" and the "spatialization of time" contribute to an altered duration that extends beyond the screen to influence the "sensory, perceptive and proprioceptive activities" of those persons caught up in the varied tempos of a technological environment.

Chapter Four:
The Social Consequences

The first three chapters explained the phenomenological method, defined the communication matrix, detailed the transition from orality to print, and explored the unique features of the screen. Drawing on the concepts presented in those earlier chapters, it is now possible to detail the ways in which the screen contributes to a changing concept of the self. In the next three chapters I will investigate the social consequences of the screen, offer a brief history of the American concept of individualism, and explore an alternative concept of the self.

The task is to understand how the perceptual gestalt of the screen is related to the social domain. It is necessary to return to the four points mentioned in the section on method in Chapter One; these abbreviated points constitute the framework that will guide our inquiry. The four points are: First, one's knowledge of the world is grounded in perception. Second, perceptual phenomena are linked in significant ways to one's understanding of the social world. Third, the screen alters perception, which, in turn, impacts the communication matrix. And fourth, the altering of that matrix will have consequences for our shared understanding of the social world. A careful discussion of each of these points will prepare for the work that is to follow.

The cacophony of a highly mediated world is enticing and hypnotic, but it is also personal insofar as that cacophony is experienced by a singular individual. The phenomenological claim that there is a foundation and that one can know that foundation through the lived body appears to relate to individual experience more directly than the social domain. Sight, touch, and sound are elements of one's individ-

ual experience. Such perceptual processes also sustain the social domain. The discussion of the four claims that follows challenges, in a
step-by-step process, the conceptual divide that separates individual
and social domains.

There is the social world, which can be understood in terms of
shared values, beliefs, and customs. There is the personal domain,
which is—and, curiously, is not—part of that social world. Evidenced
in our ways of thinking is a tendency to attribute experience either to
the singular individual or to a larger social order. Yet it is possible to
grasp, through inference and intuition, the ways in which individual
and collective experiences are deeply intertwined. As long as the dichotomy of individual and social domains remains unchallenged, the
phenomenological foundation is not capable of addressing a social
world that resides beyond one's individual experience. Nor would we
be capable of understanding how one's individual experience of the
screen alters the social domain.

The Four Claims

The first claim states that one's knowledge of the world is grounded in
perception. It is possible to view this claim as a statement of faith, if
not a statement of fact. Presuming one accepts that there is a real
world to be understood, it is self-evident that perception grounds
one's experience of that world. A straightforward analysis can reveal
the ways in which sight, sound, and touch confirm one's lived experience of the world. Understood in this way, the first claim is also a
statement about personal experience. I know the world first and
foremost through the human sensorium, but insofar as that knowledge is personal, it cannot necessarily be confirmed by others. To
avoid the conundrum that surfaces next, one must stop here; one
must not consider the role of social custom and convention.

One could argue that individual perceptual experiences are largely
shaped through communication with others. Wittgenstein's later work
details the ways in which shared language games shape our perception of everything from color to pain. We learn the categories that define color, shape, and distance, for example, through reciprocal
interactions with others. Those categories, in turn, influence how one

perceives color, shape, and distance. The promise and the peril of phenomenology reside in the claim that there are legitimate forms of knowledge prior to such conceptual processes and, furthermore, that such ways of knowing substantiate, or at the very least confirm, linguistic processes.

Wanting to understand the full power of the first claim, we are inevitably faced with a dilemma. Introduce shared customs, and it becomes difficult to identify which elements reside prior to such social constructions. The roles of custom and convention, properly understood, can make it difficult to recognize the primacy of perception. Omit such shared customs, and the claim that perception grounds one's awareness is restricted to individual experience. Any mediation of one's perceptual experience—for example, by print, screen, or telephone—will also be viewed as an individual rather than social experience. Insofar as the screen alters one's perception, such shifts could be understood as merely a matter of personal awareness. The focus on individual experience inevitably prevents one from grasping the broader social implications of mediation.

The second claim addresses this dilemma straightforwardly by stating that one's perception contributes in significant ways to an understanding of the social world. This claim challenges tacit assumptions about the dichotomy of individual and social experience. One's perception, which might in other contexts be viewed as merely anecdotal, now provides evidence of something larger than individual experience. The question is, how exactly is this possible?

The social sciences deal with the divide between individual experience and the social domain through the use of quantitative methods. Probability theory posits a degree of certainty arising from large numbers. The dilemma of the individual perspective can be usurped by lumping enough individuals together to make a credible group. The difficulty with probability theory, as qualitative researchers recognize, is that some phenomena are more amenable to numerical analysis than others. The divide between singular, sometimes idiosyncratic details, and that which can be quantifiable, challenges those who are not willing to reduce experience to quantifiable data nor to assume that individual experience is merely anecdotal.

The challenge is to construct a bridge between individual entities and larger processes. No discipline can escape this dilemma, and each academic specialty deals with this challenge in different ways. Consider the way in which English literature champions the personal circumstances of particular lives in order to create a broader frame of reference for the social concerns of a historical era. In turn, mathematics investigates complex patterns, numerical and symbolic, to discern the pulse of everything from cellular division to chaos theory. If philosophy, history, or psychology matter, it is because each of these disciplines has succeeded in constructing a bridge from the particulars to that which can be generalized.

People who pursue different disciplines hold radically diverse opinions about how such a bridge should work, hence the divide between the humanities and the sciences, the arts and the social sciences. By stating that one's perception contributes in significant ways to an understanding of the social world, the second claim merely posits that such a bridge exists, without spelling out how that bridge might work. In order for this claim to bear weight, it is necessary to provide structural support.

An example, drawn from the previous chapter, may help to illustrate this point. Chapter Three focused on the unique features of the screen but made no attempt to distinguish between individual and shared perspectives. Consider which of the following elements of the screen are features of individual awareness and which are elements of the social domain: in brief, there was the discussion of the role of motion as a unique feature of the screen; an investigation of shifting gestalts; a description of the ways in which that shifting gestalt foregrounds certain elements over others; and an analysis of changing space conceptions, with references to the Inuit, modern art, and the changing urban landscape.

Some of these elements are clearly more a matter of individual awareness. To the extent that the screen reveals a world in motion, the individual viewer will experience that kinesthetic motion in a visceral way with or without articulate knowledge of the consequences. Such interaction with the screen is personal. It is my stomach that twists when the mountain climber falls from his precarious perch; I am the one drawn deeply into the scene when the camera reveals the movement of sunlight at dawn.

By contrast, the idea of a space conception emerges from a collective awareness of the broader community as evidenced through the art and science of a particular culture. The Inuit display a different orientation to space than do urban Americans. Space conceptions change very slowly, are often only tacitly recognized, and are not reducible to individual expression. That leaves the shifting gestalt, which is more difficult to identify as either an individual experience or as a shared phenomenon. As we shall see, the shifting gestalt displays elements of both domains and perhaps provides the necessary bridge.

As long as the unique features of the screen are considered to be a matter of individual experience, it is difficult to discern the consequences for the social domain. Insofar as the screen shifts the figure-ground gestalt, altering the sensory mix in terms of what is privileged, that shifting figure-ground will alter how communities experience and think about their engagement in the world. The transition from orality to print provides examples of how changes in the forms of communication can affect the broader community. In turn, the scholarship that details the transition to print also provides evidence for the claim that one's perception contributes to an understanding of the social world.

The ability to fix words on the smooth page, to create a permanent record, slowly altered how people reflected on their inner life. As detailed in Chapter Two, in the absence of any written record, the oral community relied on the formulaic memory of the elders, fostering group cohesiveness. Print made it possible to explore one's thoughts and the thoughts of others independently of the group, which encouraged an emerging concept of individualism. Print encouraged a kind of inner reflection, which, in turn, affected one's relationship to others. The transition to print offers a model that indicates the types of questions one might ask about the screen: What does the screen foreground? How is perception altered? What are the social consequences?

Such questions presume that there is a reciprocal relationship between the perceptual processes of individuals and the resulting social domain. It may be easier to see how social customs influence perception than to grasp how perception influences the social domain. This difficulty indicates a bifurcation in the way these categories are viewed, though these categories are not unrelated. Perceptual processes can influence the social domain just as social customs influence perception. Class consciousness, stereotyping, and identity formation stand as examples of the way in which the social milieu influences how one perceives one's environment. But what might count as a reciprocal example of perceptual processes influencing the social milieu?

Our ways of thinking about these categories make this question especially difficult. Perceptual processes should not be reduced to physiological processes or to a variant of the social domain. If perception were reduced to physiology, there would be little difference between one eye and another; there would be no need to consider the shifting gestalt or the communication matrix. When perception is reduced to a variant of the social domain, the divide between the individual and the social is resolved by subsuming individual experience within the social domain. Merleau-Ponty argues that the western philosophical tradition abridged the foundational powers of the body. That denial of the primacy of perception led to the bifurcation of perceptual processes and the social domain. How one hears, sees, feels, and touches the world influences the kinds of social structures that flow out of such perceptions. My awareness of and response to the world will be very different depending on whether I live in a forest or as a member of a media-saturated community. The groups that are created in the forest and in the media-saturated community will differ as well.

The third claim states that communication technologies are altering our sense of sight, touch, and movement. This claim builds on McLuhan's point by suggesting that media not only extend but change the human sensorium. Highly mediated experiences intensify that sensorium, heightening auditory, visual, and kinesthetic elements. If the experiences of the screen were identical to one's unmediated perception of the world, it would be difficult to claim that communication technologies alter the way one experiences the world. The screen is not the first technology to alter perception. Rather, all forms of communication alter perception through mediation.

Diverse forms of communication mediate in distinctly different ways, which is why the introduction of print had consequences for everything from art and commerce to law and science. The transition from orality to print is discussed in terms of the eye and the ear. A visually dominant print culture values evidence that can be seen, knowledge that can be represented, and information that can be schematized. In such an environment, it is easy to forget that the blank spaces on the printed page are as important—sometimes more important—than the spaces filled with visible information, that what is not stated can be more important than what is spelled out. The dichotomy of eye and ear no longer suffices to explain the current hybrid, highly mediated environment that turns the eye into an ear through a blending of visual and auditory forms. Hybrid technologies provide a mix of visual, auditory, and kinesthetic tempos. One feels such resonate tempos, moves, and is moved by, a kinesthetic blend.

What exactly might it mean to turn an eye into an ear? Eyes see and ears hear, as human anatomy dictates. One's awareness of the sensorium may be changed, but fundamental physiology is not altered. Yet our views about how the human sensorium functions have changed. As detailed in Chapter Three, it is possible that the senses do not operate as separate entities, except in cases resulting from specific injury. The ear confirms what the eye sees, just as one turns instinctively to visually locate the sound beyond one's shoulder. The dizzying effect of losing one's balance makes it more difficult to see straight, just as the ear supports the eye that searches for the singing bird. As explained in Chapter Three, there is perhaps a sensory mix, rather than separate and distinct senses.

Each of the senses confirms the others in ordinary and subtle ways. As orality gradually gave way to print, the dominance of the eye and a fascination for all things visual emerged. That fascination for visual forms is with us still, but we are no longer confined to the fixed silent page. The screen links image and sound, extending the reach of both eye and ear. If it is correct that one knows the world through a sensory mix, rather than through separate and distinct senses, then heightening one sense will impact the others because they are interrelated and interdependent. Hence McLuhan's concern with sense ratios and my concern with the communication matrix. This hybrid, highly mediated environment turns eye into ear, and ear into an eye, by tightly linking visual, auditory, and kinesthetic elements. That link intensifies the interrelationship of the sensory mix. Allow me to explain by comparing nonmediated and mediated experience.

In the nonmediated world, one's eye follows the sound, or sound awakens the eye, but the senses are not scripted. The screen can script auditory and visual components to a degree that is impossible with nonmediated experience. Music does not automatically support the visual panorama of a beautiful horizon. The sounds of the world do not diminish exactly at that point when one must speak in earnest about an intimate topic. Rather, sensory awareness emerges in a sort of tacit sequence. As the eye alerts a person that he has lost his balance, awareness of the visual sense follows the kinesthetic. One's body may tremble before one feels the cold. First comes this sensory awareness, which leads tacitly to another, which leads to something else, all together confirming one's presence in the world. The ear hears and then the eye seeks confirmation. If one were to hear a loud sound and look but find nothing or tremble but not feel cold, different conclusions would follow.

Typically, one does not stop to consider how the sensory mix is fashioned, remaining unaware that one sense confirms the other in micro-sequential steps. Without a director to orchestrate the action, the world sometimes intrudes. Rather than a tight link between the senses, there is the experience of discordant sensory events. The light is too bright, the sound too loud, or the tempo too fast. One navigates all of that with little awareness of the intricate demands on the human sensorium. The communication matrix, the sensory mix that sustains awareness of the world and one's place in it, ignores the ears to privilege the movement of the hand or privileges the ears by closing the eyes. Each of us must, in subtle ways, be our own director of the scene, except for those moments when the world conspires to usurp that role as happens when the cacophony catches one unaware.

The mediation of the screen alters the communication matrix by scripting sound, image, and motion to create a seamless experience. Here the ordinary world is held at bay and cannot usurp the director. As a result, sight, sound, and motion are joined more tightly. When a speaker has something important to say, the auditory background co-operates; the panoramic view is matched with music that heightens the grandeur of the scene. Where else but on the screen would one's emotions be highlighted with an auditory component or the visual sense intensified through the movement of the camera? With both mediated and nonmediated experience, the ear confirms what the eye sees. Despite this similarity, there remain significant differences between the mediated experience of the screen and nonmediated experience. The screen alters the sensorium by linking ear, eye, and motion in ways that are not feasible with nonmediated experience. That tight link unifies the eye and ear, melds the kinesthetic with the visual and auditory, turns the eye into an ear. The viewer does not seek confirmation of the eye and ear as she would without mediation. The viewer experiences but does not orchestrate the action.

At the very least, there are three ways in which the experience of the screen differs from nonmediated experience. First, there is the joining of sight, sound, and motion. Second, there is the way in which mediation heightens sensory awareness. And third, there is the experience of an altered duration. As detailed in Chapter Three, Dorfles argues that the varied tempos of our world are vastly different from the biological rhythms all living creatures experienced prior to the modern era.

Dorfles is speaking about a wide range of technologies: trains, planes, escalators, technologies that alter one's experience of space and time. Though he discusses the altered duration of film, Dorfles's point is not restricted to communication technologies. Rather, he focuses on the many ways in which different technologies alter one's somatic experience of the environment. His point is that our physiology cannot help but respond to that altered duration. Consider the tempo of the urban landscape as expressed in the configured eight-lane freeway, the visceral press of the jet on take-off, the glass elevator descending from the twentieth floor. If so many technologies in our environment create tempos at odds with biological rhythms, why the singular emphasis upon the screen?

Visualize a large city at night: the neon, traffic, pedestrians, billboards, not to mention traffic lights, towering buildings, and street lights. The pulse of the city captures Dorfles's concern with an altered duration, and that pulse arises from different technologies. Now visualize a large city at night as displayed on screen. The view is likely to be a fluid panorama of lights, form, and space as the camera pans the horizon.

Both the first-person account and the mediated experience of the city exhibit varying tempos and durations. But there is an important difference between being on the street and watching that street on the screen. It has to do with the interrelationship of sight, sound, and movement. I may be caught up in the tempo of an urban scene on a hot summer night, but that does not mean that I am able to pan the horizon, isolate the sound of a jazz player on a distant corner, or capture the scene through a quick montage of disparate views. Human eyes and ears do not operate that way without the benefit of some sort of mediation. Many technologies can alter one's sense of duration, shifting one's awareness of space and time. As McLuhan points out, the senses are extended. The screen mediates the human sensorium more directly, altering what can be seen, heard, and felt. The ubiquity of the screen makes that mediation appear commonplace when it is actually anything but common.

The fourth claim is a synthesis of the previous three and states that altering the human sensorium will have consequences for our shared understanding of the social world. Because this claim is a synthesis, it moves the conversation to a different level. The fourth claim presumes that the human sensorium can be altered and states that consequences follow from such alterations: consequences that impact individual awareness as well as the social domain. It is not just that my perception is altered or that you and perhaps others might see the world differently. The fourth claim goes beyond that to posit something else: that our shared understanding of the social world will be changed. Placed together in one statement, the words *sensorium, consequence, shared,* and *social* ricochet off one another to create oblique implications. The first three claims culminate in the fourth, which turns into a puzzle. In order to accept or reject this claim, one must understand more about how the human sensorium is altered and the nature of those consequences. These four claims, taken together, provide a framework that must be systematically detailed in order to reveal how the screen is reconfiguring the social domain.

The Communication Matrix and the Social Domain

McLuhan's use of the term *sense ratios* provides a way of understanding the matrix. McLuhan, like Ong, emphasizes the shift from the eye to the ear. Highlighting the importance of one sense over the other obscures the way in which the senses are deeply interrelated. Placing greater emphasis on the sensorium makes it possible to think of the senses as an intertwined matrix. Consider the implications of the term *matrix*, defined as an enclosure within which something originates or develops. One's experience of the world arises out of that matrix, which, in turn, is shaped by one's interaction with the world. The communication matrix takes McLuhan's concern with sense ratios one step further by suggesting that the senses constitute a matrix or ground. It is not just the eye that is privileged nor the ear that is emphasized. The experience of the screen challenges more than the ratio of the senses. The matrix is altered.

More is at stake than the amplification of eye or ear through lens or audio equipment. Though these are important developments, there is another sort of amplification at work. Nor is physiology the issue: the ear is not redesigned; the eye is not altered. Think of resonance rather than amplification. With the screen there is a shift in the way the senses reciprocate. A new resonance is created. The term *resonance* emphasizes the interdependence of the sensory mix, the way in which the eye becomes an ear. The individual awareness of how eye and ear together confirm one's presence in the world, of the tactile and kinesthetic qualities of everyday experience, all of this is altered in tacit and subliminal ways.

The connection between the communication matrix and the social domain can best be explained by considering the shift in the figure-ground that was detailed in Chapter Three. When an emerging communication technology shifts the figure-ground in a way that privileges new forms of experience—such as the shift from the fixed space of the printed page to the kinesthetic motion of the screen—and when that shift in the figure-ground is widely shared throughout a community, it will inevitably influence community life.

I stand in some fundamental relationship to the lived world as I also stand in some relationship to a community of others who share and support my awareness of that world. There is the mystery of the critical mass. Alter the experience of enough people, and the larger community will also be changed. This is the story of print as Havelock, Goody, Ong, and others detail. Why would the same process not apply to future generations of communication technology? The point that different forms of communication influence how a community interacts is not new.

A detailed understanding of how that process works with the screen as opposed to other media is essential. There are, of course, different kinds of screens, just as there are different kinds of books. The monitor displaying text is very different from the wide screen displaying digital video. The print dictionary offers a profoundly different experience than a printed page of poetry, with its lines cascading in free form. Yet print as a medium displays essential features, just as the screen displays unique elements.

Print influenced social practices and customs; the mediation of the screen also influences social practices and customs. As detailed earlier in this chapter, the second claim states that one's perception contributes in significant ways to an understanding of the social domain. In turn, altering one's perception through mediation will have consequences for the social domain. To the extent that an emerging technology foregrounds or privileges new forms of experience, the viewer, reader, or perceiver will be encouraged to see the world differently. It is possible to recognize this point and yet fail to note the broader consequences.

The way information is presented on the screen provides amazing pictures of the world. The difficulty arises in explaining how the screen, in its varied forms, is much more than mere pictures. The way in which the screen privileges motion is fundamentally different from the way in which the book privileges space. The shift in the gestalt awakens both new forms of awareness and new forms of expression. It is not feasible to think that such changes would happen only at the individual level. Each of us exists as an individual and as part of a social milieu. The pervasiveness of the screen, the way in which that screen foregrounds motion, the way information is presented: all of this enlarges awareness of the reciprocal nature of individual and social domains.

The connection between individual perception, the mediation of that perception, and the social domain can be detailed through the ways communities experience and express their understanding of space. The space conception of a particular community is related to how an individual experiences her relationship to a perceptual horizon. People organize their experiences in terms of space and time. Of course, people also use sound, linguistics, tempo, and aesthetic objects, among other things, to organize their personal and social world. But space is the basic form through which other forms are displayed. According to Edward Hall, "Man's feeling about being properly oriented in space runs deep. Such knowledge is ultimately linked to survival and sanity" (1966, p. 105). To speak is to speak about things in some order, which implies a concept of space.

Language can imply diverse spatial patterns—for example, a tangled web. To organize a social group is to impose a spatial pattern on that group through the relationship of subgroups. One's perception of social structures, of cities and architecture, as well as ideas and abstract designs, can be captured and displayed through spatial form. Everything from the narrative structure of storytelling to the question of how a community conceptualizes the relationship of ideas, histories, and public policy can be understood in terms of spatial structures. Imagine an individual with an understanding of space (and time) that differs radically from the space conception of his community. Odds are, he will have trouble communicating or being accepted as a member of the group. Undoubtedly efforts would be made to "set him straight," which is itself a reference to space.

Different technologies can influence how individuals and larger groups experience space. Collective and individual experiences of space are quite different at the beginning of the twenty-first century than at the beginning of the twentieth century as evidenced in conceptual, artistic, and scientific records. The book contributes to a concept of space that is quite different from the space conception evident with the current communication environment. As McLuhan explains, print encourages a conception of space where things are logically oriented according to visible boundaries. The book, and film as it borrowed the narrative of the book, communicates a rational, relational, and grounded space. According to Hall, "to hold space static and organize the elements of space so as to be viewed from a single point was in reality to treat three-dimensional space in a *two-dimensional manner*" (1966, p. 86). The terms *linear* and *dynamic* have been used to distinguish the space of the book from the space of the screen. Though it is obvious that the space of the screen differs from the space of the book, that difference is not fully captured by the terms linear and dynamic. The screen may display elements of linearity just as the book may display dynamic elements.

The similarities of the book and screen stop there. Film links sound, sight, and movement to foreground motion. With the development of varied screens, that display of motion becomes a dominant feature of the social environment. For a highly literate culture schooled in the book, the rapid evolution of the screen presents a unique challenge. The screen reinvents spatial relationships, challenging print-dominant assumptions about everything from narrative to temporal sequence to rationality. The logic print imposes is subverted by a form of communication that can flow with, through, and into a three-dimensional space that is nothing like the space of the book.

My point is not related to the fact that the sequence of letters on the page creates a linear order. The fact that letters follow one after the other with print does not result in a necessary linearity. The reader takes the letters as words and sentences, fashioning those into paragraphs in order to grasp the meaning. The fact that one letter follows another does not significantly influence how one thinks through the descriptions and ideas revealed through the print on the page. The fact that one page follows another, that the page length can influence the length of paragraphs, that paragraphs often result in chapters, all this speaks to the implied spatial constraints of the book far better than the fact that one letter follows another.

The screen changes one's perspective and, in turn, alters the perspective of the community. The analogy of different types of highways may help to illustrate the point. Compare the two-lane winding country road to the eight-lane highway. How one drives depends a lot on the road. Behaviors that would be appropriate for the winding country road are inappropriate on the eight-lane highway. One's experience of the world differs significantly as well. It is more than a matter of having time to pause in one environment and being caught up in traffic in the second environment. Everything—tempo, sound, speed, vision—is different on the highway; the world is not the same, and one's relationship to the horizon and one's awareness of place, space, and sequence are inevitably altered.

That is easy to observe on an individual basis. The dilemma is grasping how the perspective of that individual or of several such individuals contributes to the social domain. We know that eight-lane highways sometimes create traffic jams. It is a factor of numbers. Place enough cars on the highway, and inevitably a traffic jam will occur. Setting aside obvious causes such as accidents and road construction, it is difficult to predict exactly when, where, or even why such traffic jams occur. Yet somehow the actions of individuals on that highway contribute to the larger pattern of traffic with unpredictable results. There exist reciprocal relationships between the cars. Someone slows unexpectedly, creating a bottleneck, and as others try to avoid the bottleneck, the traffic patterns shift again. Drivers on the highway can be viewed as either individuals or an interrelated group.

On the eight-lane highway, understanding the dynamics of the group is crucial for individual safety. Media are like highways, with different patterns contributing to the flow of information. Behaviors that are appropriate for the book are not appropriate or even possible for the screen. People who use media can be viewed as either individuals or as an interrelated group of users. Understanding the dynamics of the group, whether that group constitutes readers, viewers, or drivers, is crucial to an understanding of the medium. But how does the analogy hold with reference to traffic jams? Do different media create "traffic jams"?

Consider the process of merging from a two-lane road to an interstate highway. When a community moves from a print-dominant culture to a screen-dominant culture, the flow of information, the "traffic pattern" of that community, is altered. That creates an understandable type of "traffic jam" as evidenced in the sorts of discussions that arose around television. Some argued that television was mindless insofar as the medium was nothing like books. This type of bottleneck arises when the transition from one dominant medium to another creates confusions. Few could have foreseen, at an earlier time, that television was but one part of a larger evolution of the screen.

But what of traffic jams within a particular medium, say the screen? Is there anything that is analogous to the bottleneck that randomly occurs during rush hour on the eight-lane? With the traffic jam on the highway, we can see the ways in which individuals contribute to the larger interrelated group. Relating the traffic jam of highways to the "traffic jam" of the screen might clarify how the altering of the communication matrix on the individual level also impacts the social domain. If bottlenecks are like confusions (e.g., television is mindless), then it is clear that the screen has created its own kind of confusions. A few that come readily to mind include: shared attitudes about how the screen portrays reality, the belief that the screen creates a simulacrum, fascinations with the portrayal of the world on the screen, identification with the images of the screen, and the omnipresent use of the screen in public places. Each of these examples exemplifies social, rather than individual, appropriations of the screen.

The screen has created other, deeper kinds of spatial and temporal confusions or alterations in the flow and sequence of experience (or cars, if one is thinking traffic jams). The social domain is expressed through temporal and spatial patterns. The screen alters those patterns just as the eight-lane highway altered what was possible in terms of traffic patterns. The individual experiences the altered duration of the screen, and the larger community displays the consequences of that altered duration. Spatial and temporal frameworks are challenged. The communication matrix is altered. The world is suddenly too near, too fast; past and future no longer serve as the bookmark of collective experience. Events proceed with a life of their own; space concedes multiple vantage points. Ambiguity is heightened as foreground and background shift.

People still fall in love; babies continue to be born. Geography has not changed; we understand that east is east and west is west. But we sense that something is different and have trouble explaining what that might be. This shift in awareness has not emerged suddenly; it has been building for most of the twentieth century. T. S. Eliot wrote of it, and Yeats alludes to it in "The Second Coming" at the time that film was first being developed. The screen clarifies and enhances what is already evident in other domains. There is the subtle awareness that the terms *here* and *there* do not work as they once did, as if collective reference points have shifted. It is easy to overlook, to simply continue on to one's destination, to forget what struck one as odd in transit. There is no secure perspective, no privileged position, from which to observe such changes. We are caught up in the flow, like so many cars on the highway weaving trails of mysterious lights in the dark.

To the extent that the individual lives as part of the social milieu, the individual experience of an altered duration is reaffirmed through social practices. One both experiences the world through the screen and lives in a community of people accustomed to varied screens. The broader community responds with new forms of social organization. Historically, American culture has treated individual and social domains as separate categories. The rights of the individual and the needs of the group often stand in opposition. The screen challenges that divide. The transition from orality to print created a very different scenario. With the emergence of print, it became necessary to distinguish between printers, publishers, editors, writers, and readers. In an earlier time, authors did not sign or necessarily own their words. With print it became necessary to acknowledge individual rights as expressed through copyright law. These examples speak to the social customs that supported the press. In the larger social milieu there were also changes, as new readerships and new topics were created. It was possible to explore and reflect on inner experience as never before. Print contributed to a growing awareness of individuality.

The consequences of the screen are nothing like the book. McLuhan attempted to capture the emergence of a new type of individuality in the 1960s when he referred to the cool corporate man, foreseeing a fascination with roles rather than jobs and emphasizing involvement in depth. He argues that the electric environment created a type of group identity that was tribal rather than individualistic. It was his intention to contrast the influences of print with the impact of instantaneous electric circuitry, and the screen was one element in that electric circuitry. McLuhan's emphasis on the cool corporate image and depth involvement alludes to changes that were evident earlier in the twentieth century but not fully understood.

Just as the introduction of books at an earlier time altered social structures, the evolution of the screen is altering the role of the self in the social world. Whereas print contributed to the concept of individuality, the screen challenges our assumptions about individuality in more than one way. Not only does the screen reveal the complexities of peoples and places once beyond our reach, it alters our phenomenological awareness of how we are situated in the world. In altering the horizon, the screen also alters my relationship to that horizon. One's awareness of tempo and duration, of place and space, is tacitly transformed. The syntax of the printed word is nothing like the syntax of motion displayed on the screen. The screen reveals elements the book cannot and, in doing so, creates an altered duration that leaves its mark. The book affirms boundaries, whereas the screen dissolves boundaries. The book sequences information in a fixed format, whereas the screen sequences information in a fluid format. The motion of the screen awakens in us an awareness of space and time that is antithetical to the book.

Variable-Flex Space

We are no longer a print-dominant culture, though print is certainly important. Nor are we a visual culture if by visual one means the dominance of the eye. American culture has created a society of motion that includes visual, auditory, and kinesthetic components. Our social organizations are influenced by a variable-flex space, by an altered duration that emerged in the second half of the twentieth century, was intensified by the screen, and will define the twenty-first century in ways that are difficult to imagine. Though there is no one element that captures the complexity of American culture, it is nevertheless possible to discern dominant forms and to understand how such dominant forms influence social patterns.

A community characterized by variable-flex space experiences the world differently than a print-dominant community. To experience spatial and temporal forms as a variable-flex space is to recognize the fluid interrelated possibilities of space and time. A variable-flex space is configural. McLuhan uses the term *mosaic* to capture the varied patterns created by the flow of information. A mosaic is not fixed or bounded: think of the patterns created by a group of dancers, by wind on water, by taillights on the highway, by the reflections of the city caught in the plate-glass window as you pass. Each of these examples foregrounds not vision but motion.

The screen intensifies individual and collective awareness of such motion. The terms *linearity* and *nonlinearity* do not adequately capture this crucial feature of motion. It is not just that such spatial patterns emphasize nonlinearity. Rather, a variable-flex space emphasizes relatedness. Such a space conception is not fixed, static, or predictable. Motion heightens awareness of potential relationships. Remember a favorite commercial, a scene from a movie, or a digitized video clip where the moment is caught in slow motion to become a surreal revelation of visual, auditory, and kinesthetic form. Or perhaps remember a particular scene that was played fast forward, the action a blur of form, color, and sound, the rush captivating and illusive. The significance of such displays extends beyond mere visual effects.

First of all, the effect is defined by a kinesthetic quality; the scene moves in a way that is not feasible without the benefit of a technological interface. The viewer participates in the scene, connects sound, movement and image, makes sense of the forms, and in doing so participates in an altered duration. The further evolution of the screen has heightened this ability to reveal the world through an altered space and time. The screen introduces a variable-flex space that was possible, but not readily visible, prior to such mediation. Though poets write of it, dancers explore it, and painters strive to capture it, the screen displays that variable-flex space with an omnipresence that would be overwhelming but for the fact that we are largely oblivious to the consequences.

To exist is to be situated in time and space. To be aware of one's self is to stand in some relationship to others in the world. Such consciousness of boundaries is expressed in temporal and spatial terms. This applies to the individual, to the social domain, to the physical environment, and to the psychological expression of that environment. Anyone who has stood in a darkened field at night and felt amazement at the firmament above understands the point. The self exists in some relationship to a horizon; that relationship will be defined by space and time. Alter the boundaries of that horizon, change the meaning of "near" and "far," and the self is changed as well.

By privileging perception, phenomenology offers a unique perspective on such boundaries. Phenomenology advocates that individual perception, the way in which one's eye and hand converse with the phenomenological world, creates a path that indicates the ways in which one is both situated in and related to the larger world. Mediation tacitly challenges how the eye and hand converse with the world. Mediation also alters one's place in the world, which, in turn, alters one's relationship to others. This is true of both the book and the screen.

Imagine a community that experiences space and time as something that can be known according to logical rules, a spatial and temporal sequence that is both predictable and reasonable, a place where beginning, middle, and end follow in an orderly fashion. To the extent that such a community shared this space conception, they would also share a view of the self that confirmed what was believed to be that natural progression of space and time. The bounded properties of space would also characterize the bounded properties of a self that existed in relationship to that space.

By contrast, imagine a community that conceptualized not a sequential progression but a varied-flex space. That community's view of the horizon, of other persons, of "near" and "far," would be quite different. Sequence would give way to multiplicity. Their experience would be configural; they would live in a mosaic; their awareness of relationships would be heightened. Postmodernism endorses such a multiplicity with reference to the self, to truth, and to the narratives that explain self and truth. The screen also emphasizes a multiplicity of viewpoints. Yet, as we shall see in the remaining chapters, multiplicity is not the whole answer, at least with reference to the self.

It is easy to note the way in which electronic images from around the globe make the world seem smaller. It is much more difficult to discern the ways in which the screen alters the rhythms and tempos of daily life, to sense the ways in which that altered duration is internalized in one's immediate environment. The screen reinforces a memory of spatial and temporal frameworks that are unlike anything experienced without a technological interface. The screen confounds our customary use of "near" and "far" as instantaneous communication dissolves borders and boundaries. The screen subverts implicit assumptions about time and space by revealing a world that appears to exist at our fingertips and yet is beyond our reach. The screen displays a world of varied faces and voices to which one inevitably feels some connection, though it is quite difficult to detail the specifics of that connection. The individual internalizes that altered duration; the broader community expresses it. The individual may be fascinated by the flow of sound, movement, and images, but the consequences transcend individual fascinations.

The fourth claim detailed at the beginning of this chapter states that our shared understanding of the social world will be changed. This claim can be addressed by detailing the way in which the screen alters political, economic, cultural, or aesthetic forms. It is also well documented that the emergence of the printing press and the widespread dissemination of printed material had a profound impact on political, economic, and cultural forms. Both the book and the screen alter something else, something more primary than politics, economics, or aesthetics. First the book, and now the screen, alters how we collectively think about and experience the self in the social world.

The varied screens of the later twentieth century have inadvertently contributed to a new concept of the self. This is an inadvertent consequence because no one foresaw it; no one intended to use the screen to alter the relationship of self and other. Phenomenology begins with perception and, in doing so, provides the starting point for understanding how the screen alters the relationship of self and other. Rather than being concerned with stereotypes, computer games, and sound bites, let us understand the emergence of a variable-flex space that inevitably changes one's relationship to the social milieu.

It is time to leave the phenomenological investigation of the screen. Chapter Five will present a sketch of the emergence of individuality in America. The concept of the self that emerged in full form with a print-dominant culture must be detailed in order to provide a reference point for the discussions to follow in the final chapter. The pervasive concept of individual autonomy, with its concordant associations, will be compared to the concept of the self that emerges with the dominance of the screen. In order to explain this new concept of the self, it is necessary to narrow the discussion to particular elements: the concept of individual autonomy, inner and outer experience, public and private domains. Like Chapter Two, which provided a sketch of the transition from orality to print, Chapter Five establishes the framework necessary to apply the insights gained from a phenomenological investigation of the screen to the social domain.

Chapter Five:
The Self in the Social World

We turn now from an exploration of the screen to an investigation of American individualism. Communication technologies are altering how Americans experience and think about the self in the social world. As we shall see, the concept of individualism is neither fixed nor absolute; there are many ways to approach a discussion of the self. This text addresses the relationship of the individual to the larger group as evidenced in American political and legal thought. The very idea of democracy, as the nineteenth-century French statesman Alexis de Tocqueville was aware, embodies not only inalienable rights but inalienable tensions between the individual and the larger group. In *Democracy in America* (1956), originally published in 1835, Tocqueville envisions the individual to be relatively helpless and in need of public association:

> If each citizen did not learn, in proportion as he individually becomes more feeble, and consequently more incapable of preserving his freedom single-handed, to combine with his fellow-citizens for the purpose of defending it, it is clear that tyranny would unavoidably increase together with equality. (p. 198)

Such tensions are played out in our collective understanding of what it means to be an individual. As we will see in later sections, there are different ways of characterizing American individualism just as there are different responses to the tensions inherent in that concept.

The First Amendment identifies, in abbreviated form, the rights of the individual as well as indicates a relationship between the individual and the larger public. The terms *individualism, public, private, inner,* and *outer* identify the ways in which we collectively think and speak about the relationship of self and other. These terms cannot be separated from a complex American perspective that holds certain truths to be self-evident: truths that are captured in law, debated in public forums, and invoked in public policy decisions. The American concept of the self embodies deeply-held conventions concerning human rights, the powers of the state, and the expression of personal liberty. For Americans, the self is an essential concept that embodies a code of conduct.

The screen is changing the way we collectively think about and experience the self. The print revolution provides an appropriate precedent for this claim. As explained in earlier chapters, the book led to a new way of thinking about the self. Though it would be simplistic to presume that the development of the printing press was the single cause, many scholars agree that print contributed to the emergence of individualism. If a previous technology contributed to a changing view of the self, it is possible that today another technology can reconfigure the divide that separates private and public life.

The idea of the self is pivotal, turning both outward to address the community and inward to confirm one's private experience. The American concept of individualism, as expressed in its varied forms, requires the fundamental distinction of inner and outer experience. By definition, the concept of individualism characterizes the way in which the self establishes an identity that usurps or transcends the constraints of community life. The individual privileges a domain of experience that cannot be defined solely in terms of the group. The boundary of inner and outer experience can also be addressed in terms of private and public experience. This chapter will discuss variations on the theme of individualism, from the transcendental to the postmodern, in order to reveal how the screen reconfigures that boundary.

In *The Lonely Crowd* (1961), David Riesman addresses the mid-twentieth-century fascination with the shifting boundary between the private awareness of self and the public expression of identity. Distinguishing between inner- and outer-directed personalities, Riesman details the tensions that have characterized these domains for Americans. He argues that "in western history the society that emerged with the Renaissance and Reformation and that is only now vanishing serves to illustrate the type of society in which inner-direction is the principal mode of securing conformity" (p. 14). Inner-directed man is identified as one who can live socially without the strict and self-evident support of traditional community life (p. 14). Riesman links the rise of the inner-directed personality with print, among other factors: "[Print] can take the process of socialization out of the communal chimney corner of the era depending on traditional-direction and penetrate into the private bedrooms and libraries of the rising middle class" (p. 89). If the inner-directed man was vanishing in the twentieth century, how would one characterize the self at the beginning of the twenty-first century?

Riesman's response is to argue for the autonomous self. In doing so, he identifies the need to integrate inner and outer experience. The autonomous self is one "who on the whole [is] capable of conforming to the behavioral norms of [one's] society...but [is] free to choose whether to conform or not" (p. 242). For Riesman, the resources of character suggested at least the possibility of an organic development of autonomy out of other-direction.

McLuhan was also interested in the organic experience of the self, in a synthesis of awareness that transcends the structures imposed by a print-dominant culture. Both Riesman and McLuhan, writing in the 1950s and 1960s, were interested in defining the relationship between the self and the social world. While Riesman argues for the autonomous self, McLuhan, as indicated in Chapter Four, advocates the outer-directed, corporate, group experience. Despite their differing responses, it is noteworthy that both scholars discerned that the self in the social world—whether defined as inner and outer, or private and public—was undergoing a transformation. Both sought to explain a type of emergent social identity. That desire for integration, as expressed in the 1960s, evolved by the end of the century into the question of whether integration was even possible, as postmodernists argued that fragmentation was the defining motif.

The legal distinctions regarding privacy offer yet another way of identifying what belongs to the person and what resides within the public domain. Just as our culture distinguishes between inner and outer experience, we also distinguish between public and private domains, applying different legal standards to each. The complexity of such standards speaks to the ways in which Americans struggle with these concepts. As evidenced in libel law, different standards of care apply, based on whether the defendant is recognized as a private citizen or a public figure. Though every person is thought to be capable of inner experience, public figures or officials may not be granted equal legal rights to privacy.

Inner experience is not quite the same thing as privacy though the colloquial use of these terms may blur such distinctions. The term *privacy* indicates more than a mere legal distinction. Americans cherish their privacy and strive to protect it from the demands of both workplace and marketplace. The terms *inner* and *private* designate a type of experience that is separate from the social domain. In oral societies, the distinction between private and public life was not needed. Print contributed not only to the awareness of a need for privacy but to heightened awareness of inner experience as journals, autobiographies, and personal letters introduced private experiences into the public forum.

The terms *privacy* and *individuality*, as well as the concept of inner experience, share a family resemblance. These terms are related: each indicates a slightly different way of negotiating that boundary between the self and the social world. Shift the way one term is used, and the meaning of the other terms will be altered as well. To be an individual requires a degree of privacy. In order for the concept of privacy to be possible, one must understand the concept of inner experience. At first glance, it appears that the screen alters how one thinks about privacy by providing a forum for topics that were at an earlier time considered taboo. From talk shows to Internet chat rooms, discussions of private affairs flourish. But privacy is undermined in more ways than through the content of the screen. Alter one's experience of the world, and inevitably one's awareness of self and other will be altered as well.

The screen alters the boundary between inner and outer experience, challenging the distinctions of private and public. The intimate discussions of the screen do not alter that boundary; the content of the screen does not reveal the true power of this technology. The screen challenges the distinctions of private and public life through the representation of an altered duration. What comprises the self, if not an awareness of temporal and spatial forms? It is possible that this point was implicit in Ong's use of the term *secondary orality*. Perhaps implicit, because Ong was referencing the way in which broadcast media could access a larger audience. At that point, he was not referring to the power of the screen to create an altered duration. Nor was the screen as ubiquitous then as it is now. The details of oral versus print cultures, as Ong and others offer, draw on the distinctions of public and private life. Orality has been characterized as a pervasively group experience whereas print encourages private reflection through the activities of writing and reading. Post-literate culture cannot be adequately explained by invoking the dichotomy of private and public experience. In a strange way, post-literate culture transcends the dichotomy invoked by the customary use of the terms *private* or *public*. That language fails to capture the American experience in the twenty-first century because the boundary has shifted.

Change the way a community is situated in the world, and its ways of thinking about and experiencing the self will change as well. What appears to be the most enduring of concepts—the self—might be the most vulnerable. All sorts of technologies can reconfigure a community's awareness of its place in the world. This is true of submarines, roller skates, airplanes, paintbrushes, even clarinets. In turn, the American heritage of frontiers and innovation has encouraged an enduring, expansive view of the self, suggesting limitless potential.

The screen changes none of that. The screen simply alters one's perception of space, time, and duration. It seems merely an individual, private, somewhat idiosyncratic experience, but for the fact that one's perception of space, time, and duration constitutes the fabric of which one's experience of the world is constituted. Submarines, roller skates, airplanes, paintbrushes, and clarinets can be used, set aside, taken up, or climbed out of. Their effects are not as pervasive. These tools, if we may call them that, are not ubiquitous.

The screen, by contrast, is an extension of our nervous system as McLuhan pointed out in a prescient sort of way. Not one individual's nervous system but a collective, shared nervous system: a network. Though the paintbrush will shape the artist's awareness, and the instrument will shape that of the musician, the community has but a limited number of such creative types. Almost everyone in that community will have a screen. Few of us have failed to notice the montage of a world pixeled and spliced, cut and pasted, zoomed, panned, and digitized. Compared to a submarine or a clarinet, the screen appears to be passive. We gaze at it rather than using it in an explicit way to do, make, or create something. In fact, it is recreating us by revealing the mysterious ways in which our view of the world contributes to our view of the self.

The Concept of Individualism

The concept of individualism—that the individual is to be privileged in some way—is a fairly recent one. The meaning of this concept is subject to change despite the fact that American culture and customs depend, in subtle and profound ways, on this view of the self. The concept of individualism has informed diverse practices, from the rituals of gift giving, to the structure of social organizations, to the pursuit of science and statutes of law. Insofar as individualism reveals the bedrock of American conventions, changes in how we think about the self have far-reaching implications.

Because the concept of individualism captures that which is both foundational and taken for granted, it is crucial, and crucially difficult, to note the fault lines, to discern the ways in which the landscape shifts over time. Consider how the open prairies of a vast continent have been gradually superseded by, among other things, telecommunication networks that reconfigure space and time. In turn, shifting concepts of the self evolve in response to new forms of connection.

As detailed in Charles Taylor's *Sources of the Self* (1989), the modern concept of the self embodies a moral vision. The self embodies not only the geography of the soul, as transcendentalists might have argued, but the map that "shape[s] our philosophical thought, our epistemology and our philosophy of language, largely without our awareness" (p. ix). In privileging the person, individualism also defines the relationship between the person and the broader community. The terms *private* and *public, inner* and *outer* constitute but a few of the distinctions that raise epistemological and ethical questions. According to Taylor:

> By the turn of the eighteenth century, something recognizably like the modern self is in process of constitution, at least among the social and spiritual elites of northwestern Europe and its American offshoots. It holds together, sometimes uneasily, two kinds of radical reflexivity and hence inwardness, both from the Augustinian heritage, forms of self-exploration and forms of self-control. (p. 185)

The modern self demands such inwardness, and yet, as a form of experience belonging solely to the private person, inwardness requires a counterpart. Inwardness must be juxtaposed in some way with outer experience. Each element, in turn, defines the other.

There exists an uneasy tension not only between the two types of inwardness identified above but between inner awareness and outer experience. That uneasy tension can cast the locus of the self into doubt. Is identity constituted in community or in the person? Are the desires of the heart to take precedence over the community? Can the community usurp the privacy of the person? In American culture, differing ways of navigating such tensions have contributed to the fault lines of a shifting landscape.

Taylor details the "large-scale transformation in common assumptions and sensibility" (p. 305) that emerged in eighteenth-century Europe. The result was a concept of the person that can be summarized in three ways: "it prizes autonomy; it gives an important place to self-exploration, in particular of feeling; and its visions of the good life generally involve personal commitment" (p. 305). For Taylor, the rise of this European culture "was complexly related to the changes in economic practices, administrative structures and methods, and disciplines...such as agricultural improvements, better-disciplined armies, more effective social control" (p. 306). As is always the case with complex social events, there were undoubtedly many factors that contributed to this emerging view of the self.

New social, economic, philosophical, political, and religious influences coalesced in a moral order that privileged the person in an unprecedented way. Unconstrained by the traditions of the past, America advanced this experiment through its struggle for religious and political freedom. By the nineteenth century, Tocqueville recognized that "*individualism* is a novel expression, to which a novel idea has given birth" (1956, p. 192). In the distant past, the idea of individualism would have been wholly incomprehensible rather than novel. Though it might be difficult to imagine, in the distant future this novel idea may become incomprehensible once again.

Individualism is central to the American way of life as we have known it. Taylor's summary of autonomy, self-exploration, and personal commitment as the defining features of the person sound familiar because these are the values that have significantly shaped the American view of public and private life. Drawing on the personal stories of individuals, Bellah et al.'s *Habits of the Heart* details the different types of individualism that have contributed to the broader discussion of what it means to exist as an autonomous person in community. Distinguishing between biblical, republican or civic, and mythic or rugged individualism, Bellah and his collaborators explore the ways in which "Americans use private and public life to make sense of their lives" (p. 20). *Habits of the Heart* aptly captures the different strands of American individualism as well as the gradual evolution of this concept. The balance between autonomy and commitment has never been an easy one:

> The American understanding of the autonomy of the self places the burden of one's own deepest self-definitions on one's own individual choice. For some Americans, even 150 years after Emerson wrote "Self-Reliance," tradition and a tradition-bearing community still exist. But the notion that one discovers one's deepest beliefs in, and through, tradition and community is not very congenial to Americans. Most of us imagine an autonomous self existing independently, entirely outside any tradition and community, and then perhaps choosing one. (p. 65)

Privileging individual autonomy in this way also privileges the idea of a private and separate domain that belongs to the person, a type of inner experience that offers direction and guidance. For at least one version of individuality, meaning arises not from religious authority or from the values of the community but within the solitary depths of personal autonomy.

Bellah and his collaborators argue for a return to the values of the community. They understand, as Tocqueville foresaw, that individualism can result in the isolation and eventual alienation of the person. The crucial dilemma, as Taylor points out, was that the community of the twentieth century was also bereft of meaning:

> They [Bellah and his collaborators] write as though there were not really an independent problem of the loss of meaning in our culture, as though the recovery of a Tocquevillian commitment would somehow also fully resolve our problems of meaning, of expressive unity, of the loss of substance and resonance in our man-made environment, of a disenchanted universe. A crucial area of modern search and concern has been elided. (1989, p. 509)

Taylor makes an important point about the crisis of the twentieth century. It was not clear that a return to civic life could have solved the dilemma of a growing alienation, given the loss of meaning evident in the public domain. Scholars attribute that loss of meaning to different causes. What Taylor referred to as "the loss of substance" has been discussed in terms of the failure of science to provide the necessary answers, the chaos of the Second World War, and the rise of existentialism as an expression of such concern. Yet this was not the first nor the last crisis of meaning. The modern search has proven to be one in a series of such searches. The crises of the Reformation and the Enlightenment stand as striking examples. It seems that profound change necessitates crisis. Profound change also calls for new paradigms, new ways of envisioning the place of humans in the social world.

In both Bellah's and Taylor's work, the locus of the self presumes an underlying dichotomy of self versus community. The basic landscape remains unchanged. Bellah questions how the individual is to construct a meaningful life in community. Taylor questions the implications of a civic life that fails to provide meaningful engagement. Regardless of whether the individual retreats to a solitary perspective or returns in some way to civic engagement, such choices presume a basic tension between the self and the community, between personal autonomy and the constraints of the social domain.

Whatever the specific view of individualism on which one focuses—religious, civic, expressive, mythic—the autonomy of the self is inevitably contrasted with, or juxtaposed against, the needs of the larger community. It is this very juxtaposition, according to Bellah, that requires some sort of balance for the health of both individual and broader community. Obviously, not all types of individualism are concerned with balancing the autonomy of the individual against the constraints of the community. Mythic or rugged individualism stands as an example. Nevertheless, it is this tension between the individual and the community that provides the guiding framework for the American perspective on the person. The dilemma is that this guiding framework—as exemplified by the concept of individualism—and the requisite ways of privileging the person are at odds with the emergence of a different cultural motif that is facilitated, in part, by the rapid development of communication technology.

I am speaking about an American tradition at a very compelling and specific moment in time, when American values are influencing, in both positive and negative ways, an international audience. Just as the call for basic human rights is growing internationally, the inalienable American rights of freedom, liberty, and justice are being challenged by a new environment. The concept of American individualism no longer adequately captures our collective experience of the self in the social world.

Though the concept of secondary orality offered a helpful way to understand the implications of a mass audience in the mid-twentieth century, it does not suffice to explain the current environment. No doubt, there are elements of the oral culture that can be identified in the current milieu just as there are also elements of American individualism that will endure in the coming era. Perhaps the emphasis on human rights is one such element that will survive to be integrated into a very different concept of the person. Nevertheless, individualism, as it has been discussed in American historical perspectives, will not provide an adequate way of characterizing the person by the end of the twenty-first century. By implication, related terms such as *inner* and *outer*, *private* and *public* will also become increasingly problematic. This is already evident in discussions regarding the public domain. Communication technologies have contributed to rapidly evolving technological and social environments that challenge the traditional distinctions of private and public life. As we shall see, there are also potential difficulties with the terms *inner* and *outer*.

Dichotomies of the Self

As stated earlier, the dichotomy of inner and outer is intrinsic to the concept of individualism. To the extent that the individualism concept is shifting, there will be a comparable, perhaps tacit shift in the boundary between inner and outer experience. The distinction of inner and outer may be understood as foundational to all human experience, or that dichotomy may arise in response to particular historical events. Even if it were the case that all people everywhere distinguish between inner and outer experience, it would still be necessary to spell out the permutations on this theme. For the Buddhist, all experience is inner directed. The physical and social world is only an illusion. Inner experience would play a much different role for the transcendentalist than for a member of a highly collectivistic society. The postmodernist who argues for multiple selves is positing a different view of inner experience than the modernist who argues that a return to the commitment of civic life is crucial for the individual. Each variation highlights subtly different views of the self.

Our highly mediated environment challenges the dichotomy between inner and outer experience. Though the current discussion is being played out largely in terms of private and public life, also implicit in that discussion is a concern with the nature of inner experience. Inner experience is, by definition, distinguished from the shared experiences of the community. The legal concept of privacy supports the belief in a domain of personal experience that is set apart from the constraints of the social world. If the community agrees that the details of private lives are now appropriate for public consumption, the boundary of inner experience will be increasingly called into question. Just as our experience of the world today is fundamentally different from the experiences of the eighteenth century, our shared concept of the self today certainly must be different from the concept of individualism that helped to shape the laws and customs of early America.

This complex American heritage links individualism, privacy, and freedom of speech. The relationship between these terms evolved slowly as the consequences of free speech were addressed through social and legal practices. The American experiment with freedom of speech has not yet ended. The inherent tensions between the rights of the individual and the constraints of the larger community are now being played out with a vengeance. Current concerns regarding freedom of speech reveal the conflict of old and new. Just as the original colonies struggled to define what it meant to advance freedom of speech, today we struggle to adapt earlier ways of thinking about First Amendment freedoms to a new communication environment. The original debate was motivated, in part, by the opportunities print created. The current debate is motivated partly by the opportunities the screen creates as well as the various technologies that support the screen.

Scholars have argued that the shift from an oral to a print culture encouraged the development of an individualistic perspective. As detailed in the second chapter, the traditions of orality favored the group over the individual. Without a written record of events, it was crucial to foster a collective memory that could support the practices of the community. Simply stated, first with writing and then with print, it was possible to manage collective memory in a vastly different way. Taylor's reference to "changes in economic practices, administrative structures and methods, and disciplines" (1989, p. 306) is seen by Eisenstein to be one of many significant consequences of print. With writing it was possible to keep records, write instructions, improve bookkeeping, and explain new scientific theories. With print it was possible to create exact copies and share those copies with a wider audience. While writing provided a way to examine one's inner thoughts, print made it possible to share such thoughts with a much larger audience than was possible with a limited number of hand-copied manuscripts.

Today the terms *collectivistic* and *individualistic* might be used to distinguish between oral and print cultures insofar as the former privileges the group and the latter privileges the individual. But these terms belong largely to the twentieth century. It would be impossible to consider such a distinction in an oral culture because there would have been no frame of reference for thinking about the self in any way other than as an extension of the group. The modern use of the term collectivistic posits a dualism. Properly understood, this term must be contrasted with the alternative, which is individualistic. The concept of the self in the social world as we know it today is a product of that dualism. The self in an oral community lived within a resonant spoken tradition that did not distinguish between inner and outer experience in the way that we do now. What appears to be a natural and obvious distinction is actually the product of a particular historical period, a part of the moral good that Taylor eloquently discusses. To be a person is to be autonomous, which is to recognize oneself as distinct from others: to know one's thoughts, to acknowledge the desires of one's private passions. It is probable that none of these criteria could have aptly characterized what it meant to be a person in a preliterate community. The concept of the self as we know it did not exist.

The self cannot be determined solely by communication technologies such as print any more than war between neighboring countries could be explained by the invention of a single weapon. Complex social events require complex explanations. Nevertheless, to overlook the impact of such technologies would be shortsighted. The study of anthropology details the profound and unexpected changes that occur when new tools are introduced into a community. Social structures can be influenced by such things as gunpowder, farm equipment, or writing instruments. The printed book changed the way people practiced economics, housekeeping, and science. Perhaps it is easier to explain how the book altered trades and professions than to explain how the book altered Western Europe's view of the self. Like the book, the screen is altering the way people practice economics, science, and housekeeping. In turn, the screen alters our view of the self. The changes the book created can serve as a model for understanding the implications of the screen. Drawing on the themes presented in Chapter Two, I will detail the consequences of the book in order to address those implications.

Consequences of Print

With the emergence of print came changes not only in record keeping but in attitudes about the autonomy of the individual. Such changes evolved over centuries. The Gutenberg press was invented in the fifteenth century. In the seventeenth century, John Milton made one of the first arguments for the freedom of the press (Zelezny, 1993, p. 33). According to Taylor, by the turn of the eighteenth century the modern self was emerging. The ten amendments to the American Constitution that spelled out the rights of the individual versus the state were ratified at the end of the eighteenth century.

Without belaboring the point that there were multiple factors responsible for this emerging view of the self, let me highlight the ways in which scholars have linked this social transformation to print. Havelock, Ong, Goody, Eisenstein, and others have argued that the ability to write out one's thoughts allows for a degree of reflection that was not possible in preliterate cultures. Memory is freed from the burden of keeping track of the collective history of the community. A private space is created for thoughts that originate not within the shared oral stories of the tribe but within the person. Language can be used as a tool of exploration rather than as a mnemonic device. It is now possible to keep track of one's thoughts, to review and follow those thoughts out to a new conclusion. Regardless of content, style, or grammar, the written word privileges the feelings and thoughts of the individual writer. Print also makes it possible to share one's thoughts with a wider audience. As books became widely available, the reflective process of writing was matched by the equally reflective process of reading. Both reader and writer are physically set apart from the oral practices of the community. To read and to write requires a special kind of solitude, where one is absorbed in a conversation of a very different kind. The dialogue is internalized in a way that privileges an awareness of inner life.

Public discourse, as well as inner dialogue, was changed by the power of print. Despite the best efforts of church and state, it proved increasingly difficult to silence dissenting voices. As the printing press subverted the powers of religious and governmental authority, those in power fought back to censor authors who challenged institutional views. In order to punish someone for seditious libel, it was necessary to identify the author. As writers were gradually distinguished from editors and publishers, the idea of ownership emerged. This, in turn, prepared the way for the modern concept of copyright. Freedom of speech, copyright, and privacy evolved in response to the powers of print.

Print also altered the human sensorium, hence McLuhan's phrase, an eye for an ear, and Ong's profound fascination with the distinctions of orality and literacy. Words in print replaced the resonance of the voice and, in the process, altered the balance of the senses. For the highly literate, the dominance of the eye might delegate the senses of sound, touch, and smell to a secondary status. Like the emerging concept of the modern self, this process evolved gradually. Even today, there are many cultures and communities that are neither highly literate nor visually dominant. Nevertheless, despite the diversity of many co-cultures, until recently America could be characterized as a print-dominant society. Now the screen competes with the book.

When considering the effects of print, it is easier to grasp the ways in which the solitary reflections of the writer led to a new awareness of the person than to understand how the use of a typographic alphabet encouraged awareness of individual autonomy. The internalization of the voice tells but a part of the story. The way in which the eye privileges the fixed letters on the page reveals another aspect. According to both Ong and McLuhan, text encouraged a certain type of perspective, a fixed point of view, an orderly way of delegating ideas according to the structures of sentences, paragraphs, and chapters. Of course, print comes in many styles, and as poets know, the paragraph is not the only choice. Regardless, it is appropriate to note that print fixed much more than the words on the page.

Print provided a vehicle of thought that privileged a logical approach, which some hoped might fully capture our experience of the world. With text came a worldview that designated a certain order not only for laying out one's thoughts but for identifying how one stood in relation to such thoughts. The desire for certainty, the belief in validity, the faith in rationality, all sprang to some extent from a fascination with the medium of print. It was not just that the alphabet might finally reveal the truth but that it would, in turn, provide a map for locating the geography of our thoughts, revealing that which had been previously hidden from view. The promise of print was that it might finally reveal the connections that bind question and answer together. This was the desire of poets and scientists alike. It was not to be. Instead, the medium of the book obscured the ways in which text structured the writer's thought.

The experiences of the book and the screen are profoundly different. The screen emerged not at the beginning of the modern era but at the end. The desire for an encompassing rationality, the promise of certainty, had begun to dissolve under the glare of a world both wild and unpredictable. Science unraveled the universe only to discover conundrums far greater than anyone could have envisioned. The horrors of the Second World War called everything into question. Threatened with annihilation, the human response was to question one's relationship to the cosmos and to other living creatures, to question one's relationship to the unconscious and to the natural order, to question one's place in a world that had rapidly become disorienting. From the middle of the twentieth century onward, the glowing screen was part of that process.

The book serves as a model for the screen, despite the fact that these communication technologies emerged at very different moments in history. The book changed how people thought and, in turn, influenced how they thought about themselves and others. As a medium of communication, print privileged certain forms of expression. A new medium of communication will, in turn, privilege other forms of expression.

Differing Views of Individualism

Laws concerning speech, copyright, and privacy, written in response to print, also indicate a way of thinking about relationships between the self and the larger community. The American concept of individuality—of an autonomous self with inalienable rights—is embodied in the legal traditions of the twentieth century. As Katsh explains, "at the heart of our legal system is a view that society is constituted of autonomous separate individuals and that such individuals are more important than any larger constituent group" (1989, p. 239). Just as laws created in response to print may fail to address the new digital environment of the screen, they also fail to capture an emerging concept of the self that arises, in part, as a consequence of that new communication environment. Our laws, like our conceptual language, are necessarily grounded in the traditions of the past. While such traditions help to maintain stability, they are also, by definition, unresponsive to the emerging tensions of the present.

Freedom of speech, copyright, and privacy would have been incomprehensible prior to the medium of print. In an oral community, freedom of speech was not necessary. Without written texts, the emphasis was on collective remembering. Such communities were bound by tradition. Dissenting voices could be ignored, because such dissent was not fixed in print. In turn, copyright was not needed, since ownership of creative works had not yet emerged as a concept. According to Katsh, the modern concept of privacy arose, in part, out of a desire to protect the reputation of individuals from the press. Obscenity laws arose in response to a print environment that made it possible to circulate offensive works. Laws regarding the nature of speech, constraints concerning copyright, and attitudes about obscenity also capture the larger community's view of the self. Copyright is now under attack, as software makes it possible to usurp the concept of ownership. The enforcement of obscenity law is problematic as well. In the mid-twentieth century, an era when Lenny Bruce was jailed for using explicit language in his comedy routines, who might have imagined how very difficult it would become to regulate the flow of pornography and obscenity? The Court could not have foreseen a communication environment that made it possible to send information around the world in a matter of seconds.

This country is shaped not only by the traditions of the past but by a very specific view of the individual that arose in a particular time as a consequence of particular circumstances. Not only do our laws grant certain rights to the individual, they also indicate the relationship of the self to the social world. American political and legal practices codified the inherent tensions that exist between the individual and the community. The response to those tensions, in the form of legal precedents and social customs, led to a revolutionary view of the self as separate from and independent of the community. As Tocqueville, Taylor, Bellah, Riesman and others have explained, implicit in that revolutionary view of the self was the possibility for both great alienation and great commitment. The implications of the screen are now being played out against the traditions print established.

One way of understanding that heritage is to locate it not only within the law of a particular era but within the philosophy of American transcendentalism. The writings of Ralph Waldo Emerson and Henry David Thoreau privilege the autonomy of the individual over the community. Though their views may sound somewhat extreme more than a century later, such views embodied the hopes of a people who placed their faith in the autonomy of the individual. As explained by Michael Meyer in the introduction to Thoreau's *Walden* and *Civil Disobedience*, first published in 1854 and 1849, respectively:

> Thoreau followed Emerson in locating God within one's soul and in nature. Because absolute values and authority could be discovered within one's self rather than in the pulpit, the tract, the statute book, or the marketplace, a person could be totally independent and free if this divinity was developed and given expression. (1983, p. 13)

This perspective provided one answer to the difficulty of existing as an individual within a community: cut loose the community and retreat into the sanctity of the soul. Thoreau did not place his faith in the opinions of the multitude. He privileged the self that searched for origins in another sort of landscape. His is a romantic and rugged view of individualism that disdains the constraints of community in favor of an inner direction.

Thoreau was not the first to struggle with these issues. Those who have felt the peace of the wilderness can readily understand why he might have retreated from the wearisome discourse of community life. The rugged individualism exemplified by such self-sufficiency remains enticing for many who find the social domain to be bothersome at best and a hindrance at worst. According to K. J. Weintraub's account in *The Value of the Individual* (1978), roughly less than a century before Thoreau, Johann Wolfgang von Goethe also "thought of nature as a ceaselessly pulsating reality, an order of creation constantly transforming itself; he thought of individuality as an evolving part of the natural realm" (p. 336). But Goethe, unlike Thoreau, did not believe that the individual must transcend the constraints of community: "Goethe did not set the self and the world against each other; he did not think well of self-cultivation at the cost of the world" (p. 336). He was interested in the intrinsic relationship between self and environment, whether that environment be woods or town square. In striking opposition to Thoreau, Goethe would argue "that only mankind together is the true man, and that the single individual can be joyful and happy only when it has the courage to feel itself as a part of one" (p. 376). Thoreau's writings indicate that he would have had little patience with Goethe's interest in the experience of mankind together.

The concept of the self in the eighteenth and nineteenth centuries is distinctly different from the concept of individuality operating in American culture today. We recognize the perils of rugged individualism. Thoreau's retreat could now be recognized as a form of isolation. Too often the social price of such isolation has been violence, given our immersion in a communication environment that inevitably intensifies both isolation and connection. Fax, electronic mail, instantaneous satellite communications enable us to see, feel, and hear the world in a way that would have been unthinkable for Thoreau. Even today, many advocate unplugging all electronic connections as the only reasonable response. Some argue that the traditions of civic engagement on which this country was founded have been replaced by an isolation that breeds apathy.

Our laws and traditions reflect a precarious balance between the constraints of community and the rights of the individual. Responding to that precarious balance, we have fashioned an approach to social discourse which is fraught with paradox. Individualism in all its possible manifestations requires engagement in a community and yet is threatened by that engagement. Our current concern with a growing isolation and apathy is not new. It was evident in Tocqueville's writings and continues to shape our laws and customs. Nevertheless, the isolation of Thoreau's retreat was very different from the isolation of the twenty-first century, complete with instantaneous electronic connections to the far reaches of the globe. The screen creates a very different kind of isolation.

Meyrowitz and others argue that one consequence of such instantaneous connection is the loss of community life at the local level. In *No Sense of Place* (1985), he argues that the visual images and stories of the screen usurp the connections to one's local community. By contrast, consider the isolated communities of the American frontier, without broadcast or print. News of the world beyond was delivered by the occasional visitor. In that vast frontier it could take months for a letter to arrive. The isolation of the screen is of a vastly different kind. The screen creates a new geography by virtue of a communication environment that appears to compress our experience of time and space. The images and sounds of a digitized world appear so quickly on the screen that one cannot help but marvel. Our grandchildren will not marvel. They will have internalized a very different experience of time and space. This new environment cannot be understood in terms of more or less isolation, because what we are now experiencing is an isolation of a different kind. We are aware of new ways of existing in relationship with others but less aware that the potential for such relationships can inadvertently heighten one's feelings of separation.

The claim that we are more isolated overlooks the fact that the nature of "community" has shifted radically. For the early American struggling to survive on the frontier, community meant those people one could reach by horseback. For the American of the twenty-first century, community can mean a wide range of virtual, professional, social, and local communities. Isolating oneself from community on the frontier would have been a very different activity than separating oneself from the multiplicity of social connections that characterize the digital age. Could it be that people are now both more isolated and more connected than ever before? We are not as isolated as our forefathers might have been, without telephone, daily newspapers, or radio. Nevertheless, the intuition that one's relationship to community life has shifted is correct: more is at stake than mere isolation.

Goethe provides the necessary response. This eighteenth-century German argued that "only mankind together is the true man." May I suggest that Goethe's idea of "mankind together," though certainly not intended as a statement regarding democratic participation, is nevertheless supportive of a civic community with a shared commitment to the larger good. Tocqueville wrote of the way in which "Americans of all ages, all conditions, and all dispositions, constantly form associations" (1956, p. 198). He understood the crucial role that such associations played in the democratic process. Commitment to civic life may now be problematic, in part because we are deluged daily with so many opportunities for communication. In the current environment, it is increasingly difficult to ignore those opportunities. Communication technologies, as well as changes in transportation and economics, have made it easier to express commitment to a larger cause; yet it appears that Americans are less engaged. Whereas the solitude of the frontier may have encouraged a move toward association, the deluge of opportunities today may necessitate a retreat to solitude.

A Postmodern Response

Goethe had it right, despite the fact that American transcendentalism would not have embraced his concern for togetherness. Individualism cannot exist in a vacuum. It is a fundamental American paradox that autonomy is linked to community. The more a person understands her relationship to a community, the more she can understand where she has come from, how she is different, and what is unique about her existence. To the extent that our communication technologies are altering our sense of connection to the social world, our understanding of individuality will necessarily be altered as well. Our idea of community life may be based on traditional notions of democratic participation. Nevertheless, it has been a very long time since community was defined as those people you can reach on horseback. The screen is altering how we experience community. In turn, that shifting experience of community implicitly alters one's relationship to self and other. As a consequence, the concept of individuality is being redefined.

Postmodernism offers a different response. Whereas Goethe argued that "the very specificity of an individuality is tied to the unrepeatable uniqueness of a spatial-temporal point" (Weintraub, 1978, p. 352), postmodernism questions whether the individual can necessarily be reduced to one recognizable spatial-temporal point. According to Martin Jay in *Downcast Eyes* (1993), "if postmodernism teaches anything...it is to be suspicious of single perspectives, which, like grand narratives, provide totalizing accounts of a world too complex to be reduced to a unified point of view" (p. 545). The postmodern experience of space and time evokes multiple horizons. Just as cubism of the modern era captured an object from multiple points of view, the postmodern world highlights not one secure perspective but interchangeable vantage points. This pivotal idea can be addressed in terms of physics, probability theory, and poetry as well as the visual arts, economic theories, or communication and transportation networks, to name but a few of the domains that display a shift from the certainty of modernism to the multiplicity of postmodernism.

The postmodern individual lives a de-centered existence without access to that "uniqueness of a spatial-temporal point" in which Goethe believed. Doubt becomes the first commodity; in turn, the self displays not a singular and unified identity but variations on a theme. According to Jean Baudrillard in "The Masses: The Implosion of the Social in the Media," an essay published in *Selected Writings* (1988), one result of the media environment is "a radical uncertainty as to our own desire, our own choice, our own opinion, our own will" (p. 209). Postmodernism considers identity to be a fluid creation, capable of responding to shifting social and geopolitical constraints. Because there is no privileged or fixed perspective, there is no unique self. Rather, the individual is comprised of myriad selves that emerge in response to differing social situations. According to Gergen in *The Saturated Self*, "in the postmodern world, selves may become the manifestations of relationship, thus placing relationships in the central position occupied by the individual self for the last several hundred years of Western history" (pp. 146–147). Both Goethe's and Thoreau's perspectives are overturned. To the extent that our "belief in essential selves erodes" (p. 146), the self disappears into innumerable relationships. The individual cannot discover his uniqueness through community, nor can he retreat from community to discover the inner truth of his being.

Consider the radically different views of the self espoused by transcendentalism and postmodernism. Transcendentalism advocates that the sanctity of the individual, once recognized, will deliver one to the truth. This inward turn privileges a domain of certainty which exists independently of community. Postmodernism turns transcendentalism upside down. Disparaging the possibility of truth, the individual now exists through the intertwining identities created by engaging in differing communities. Where community was once considered to be a threat to the autonomy of the person, engagement now becomes the principal way in which one creates identities. Questions of realism and relativism aside, the juxtaposition of these two views highlight the American struggle to define the role of the self in the social world. Placed side by side, they illustrate the proverbial American tension between individual autonomy and collective identity.

Both transcendentalism and postmodernism are inadequate to explain how American individualism is changing in response to our current communication environment. Transcendentalism forsakes community, and postmodernism forsakes any concept of a unique self that might exist apart from community. Each perspective denies a crucial half of the equation. Transcendentalism turns to inner experience to define the self. In this view, one must retreat from the clamor of the multitude in order to hear the truth of one's innermost being. Postmodernism turns to outer experience to define the self. Now it is the community that gives shape and form to the various identities of the person, who exists only in relationship to others. In the former, it is the individual who matters; in the latter, the community.

The problematic division of inner and outer, private and public, is embodied in these two views. Our history and traditions encourage this problematic choice. The tensions inherent in American individualism imply that one can privilege either the private experience of the individual or the public experience of the group but not both. Our ways of thinking about the self, our laws and customs, do not provide a way around this problematic dichotomy. Goethe's perspective, that "a self [is] a self only in its fruitful interplay with its world" (Weintraub, 1978, p. 362), offers a way to integrate the perilous dichotomy of inner and outer experience. But Goethe's perspective requires more elaboration, given the radical differences between eighteenth-century Germany and the highly mediated world we now inhabit.

As we shall see in the final chapter, mediation has profoundly altered the dichotomy of inner and outer experience. We will return to a discussion of the communication matrix in order to detail the consequences for the self in the social world. If the medium of print were capable of changing how humans thought of themselves, it is possible that the screen can also transform how we experience and conceptualize the self. The crucial difference is that in an earlier century, as a new country, America could escape the constraints of tradition. The same cannot be said today. To the extent that the original concept of

American individualism is being redefined, this time that transformation will surface against the powerful currents of entrenched legal and social customs. One consequence is a social world that appears slightly off-kilter. On the other hand, this historical moment offers rich opportunities. It is unlikely that the founding fathers could have grasped the broader historical implications when they laid out the Constitution and the Bill of Rights. For they were merely men, caught in a particular time and place; there was no way for them to envision the radical changes that would follow. We, too, are caught up in a great transition. It is our fate to respond to the moment, to live at the cusp of a new century, to struggle with traditions that no longer offer an adequate map for a new terrain.

Chapter Six:
Beyond Individualism

Though it was inevitable that our country would evolve into a new era, the technological changes of the last century have transformed our way of life far more rapidly than anyone could have imagined. Descriptions of those changes are often played out in the negative: such commentary grieves for what is lost without acknowledging what is gained. The violence of a new technology, or a new idea, is never simply negative. We are collectively caught up in a profound sea change. Individualism, with the concordant rights and privileges it assigns to the person, provides the key for unlocking the larger puzzle of how current communication technologies are altering the self in the social world. The concept of individualism provides the frame through which we experience private and public life. The disintegrating boundary between private and public life constitutes a sea change because that dichotomy is fundamental to the American concept of individualism. When this distinction breaks down, the crucial concept of what it means to be an individual breaks down as well. The violence that arises as a consequence of new communication technologies can best be understood through a discussion of the private and public domain because that is our heritage: that is where we began.

As Meyrowitz explains, "historians have noted the new sense of 'boundaries' dividing public and private, domestic and political, family and community in the transition between the pre-modern and modern age. In our own time, we have found a dramatic reversal of this trend" (1985, p. 314). We must grasp the implications of that reversal. Taylor alludes to such a reversal in *Sources of the Self*:

> The modernist retrieval of experience thus involves a profound breach in the received sense of identity and time, and a series of reorderings of a strange and unfamiliar kind....As a result of this, the epiphanic centre of gravity begins to be displaced from the self to the flow of experience, to new forms of unity, to language conceived in a variety of ways. (p. 465)

American individualism may not accurately capture the "profound breach in the received sense of identity and time." The self is no longer recognized as the locus or epicenter. The resulting discontinuity raises questions concerning the role of the self in the social world.

In an earlier era, Emerson privileges the individual perspective, espousing the view that "there are the voices which we hear in solitude, but they grow faint and inaudible as we enter into the world. Society everywhere is in conspiracy against the manhood of every one of its members" (1975, p. 21). By the middle of the twentieth century Carl Jung also argues that man must turn inward to gain his sense of direction. He places "the individual human being in the center as the measure of all things" (1957, p. 52). To a large degree, our shared experiences now indicate that the "measure of all things" cannot be fully captured by an individual perspective. We are all too aware that individual decisions can have long-ranging consequences for diverse groups. Public policy decisions concerning health, welfare, and education provide outstanding examples.

Technology heightens our awareness of the ways in which we are connected to others. Whether by telephone, fax, Internet, or television, we are bombarded with image, voice, sound, and movement. Meyrowitz argues that what results is a dislocation of place and a growing tendency to isolation:

> Why and how do technologies that merely create new *connections* among people and places lead to any fundamental shift in the structure of society or in social behavior? One potential answer to this question rests in the ways in which *dis*connectedness—the separation of social situations and interactions—shapes social reality. (1985, p. 23)

Yet there is also, curiously, the opportunity to listen and observe as others share their most intimate secrets. In growing more isolated, we grow more intimate. As the distinction between public and private weakens, the distinction between intimacy and alienation is also obscured. These distinctions are no longer clear, whether one is talking about electronic records, the portrayal of intimate secrets on the Internet, or the sharing of personal stories via telephone.

It is not just that it is easier to communicate but that the nature of communication has changed. It is possible to say things to a camera that one would never say to one's family in the privacy of the home. The telephone changes the nature of the conversation by omitting both the gaze of the other and nonverbal responses. Electronic mail allows us to forward messages and create conversational loops that are very different from either face-to-face conversations or traditional correspondence. Meyrowitz argues that "once widely used, electronic media may create new social environments that reshape behavior in ways that go beyond the specific products delivered" (1985, p. 15). Technology not only makes it possible to communicate in new ways; such innovations also facilitate the development of new social environments that alter our awareness of time and space as well as the role of the self in the social world.

Implicit in the writings of Ong, McLuhan, and, less directly, the work of Pierre Teilhard de Chardin are perspectives that explain some of the changes we are now experiencing. These scholars, each in his own way, are concerned with the process of social change, with the transformations of technology, with the evolution of human consciousness. In *The Future of Man*, originally published in France in 1959, Teilhard writes that "we have only to go a little further, I am convinced, and our minds, awakened at last to the existence of an added dimension, will grasp the profound identity existing between the forces of civilisation and those of evolution" (1964, p. 301). Because Ong and Teilhard, both Jesuit priests, lived for a time in the same Jesuit house in France, it is probable that Teilhard's work influenced Ong. In *Letters of Marshall McLuhan* (1987), McLuhan links his own interest in electricity to Teilhard's work: "That electromagnetism as such is an extension of the central nervous system is a persistent theme of Teilhard in his *Phenomenon of Man*" (p. 292). When placed side by side, the concerns of these three scholars prepare the way for a richer understanding of what it might mean to be an individual at the beginning of the twenty-first century.

The Evolution of Consciousness

Decades have passed since Ong and McLuhan first wrote about secondary orality and retribalization. During that time the digital divide has reconfigured communication technologies, radically altering our public and personal lives. Though these scholars agreed that new forms of communication can alter the social contract in subtle and profound ways, they were also interested in the broader question of consciousness. In *Interfaces of the Word* (1977), Ong writes that "the new medium is not just a new way of purveying what other media purvey in their own ways, but is rather the implementation of a new state of awareness" (p. 319). Ong was interested in the evolution of consciousness, arguing that such an evolution "would be impossible—unthinkable—without the alienations introduced by writing, print, and the electronic transformations of the word" (p. 47). It was Ong's task to detail the ways in which communication technologies both humanize and alienate man from himself. According to Ong, communication is always about the self:

> Communication does not have simply to do with oneself, but in fact can concern anything and everything. But it always involves the self, for communication, whatever it is concerned with, is a conscious activity, and the 'I' which I speak and which I alone can speak lies at the center of my consciousness. (p. 336)

Ong understands that this "discussion of the 'I'...would have been impossible a few centuries ago" (p. 341). The concept of an individual perspective is a fairly recent historical phenomenon, and with that concept inevitably comes isolation. To the extent that it is possible to stand apart as an individual, it is also possible to be separated from the group and, in curious ways, to be separated from oneself. This separation began with the first creation of written forms. As Taylor explains, "this freeing of nature from the iconographic tradition also carries consequences for the place of the subject. The artist who sets himself to imitate nature sees himself as standing over against the object" (1989, p. 201). As Ong explains in *Orality and Literacy*, "writing introduces division and alienation, but a higher unity as well. It intensifies the sense of self and fosters more conscious interaction between persons" (p. 179). It was the paradoxical ways in which technology altered human experience that compelled Ong to investigate orality and literacy.

As early as *The Presence of the Word* (1967), Ong was focusing on both technology and evolution. He was aware of the ways in which electronic media contributed to new forms of social awareness: "The fragmentation of consciousness initiated by the alphabet has in turn been countered by the electronic media which have made man present to himself across the globe, creating an intensity of self-possession on the part of the human race which is a new, and at times an upsetting, experience" (p. 321). That "intensity of self-possession" is the link that connects the work of Ong, McLuhan, and Teilhard.

McLuhan saw a culture caught at the breaking point, schooled in one way of thinking, even as our environment demanded a very different type of response: "We *live* mythically but continue to think fragmentarily and on single planes" (1994, p. 25). McLuhan's work can be characterized as an exploration of pattern. He grasped the way in which technologies reconfigure space, time, community, and personality. Characterizing his comments on changing media as probes, he argues that "no medium has its meaning or existence alone, but only in constant interplay with other media" (p. 26). His obsession with the medium was predicated upon an awareness of the subtle and profound environmental changes that arise with new technology. He understood that the information revolution offered new ways to configure experience. He wrote about a new social order that was emerging out of a growing awareness of interrelatedness. In *Media and the American Mind* (1982), Daniel J. Czitrom details McLuhan's argument that "new forms of electronic media seemed to have reversed the sensory fragmentation of visual space, thus foreshadowing a psychic return to the tribal situation" (p. 174). Yet such a retribalization, like secondary orality, assumes new forms in an electronic culture.

McLuhan would have agreed with Ong that the culture was moving into a type of secondary orality, though he did not use that specific term. He referred instead to *depth culture*, juxtaposing the fragmentation of a previous mechanical and specialist culture to a new experience that requires one to live with a deeper awareness of pattern, simultaneity, and interrelationship. McLuhan argues that "today it is the instant speed of electric information that, for the first time, permits easy recognition of the patterns and the formal contours of change and development" (1994, p. 352). He uses terms such as *simultaneous, mosaic, depth, interdependence*, and *interrelation of the total field* to explain this new way of being involved in the social process. One is forced to ask, how does this involvement differ from earlier types of involvement? Czitrom explains that McLuhan saw "a new sensory galaxy ushered in by electronic media that are capable of jolting our sensibilities as sharply as the printing press did earlier" (1982, p. 176). As McLuhan's work attests, though the electronic media may jolt "our sensibilities as sharply as" print, the consequences of the screen are vastly different from the consequences of the book.

To return briefly to Ong's *Orality and Literacy*, in a section titled "Consciousness and the Text," he writes that "since at least the time of Hegel, awareness has been growing that human consciousness evolves...growth in historical knowledge has made it apparent that the way in which a person feels himself or herself in the cosmos has evolved in a patterned fashion over the ages" (1982, p. 178). For Ong, this evolution depended on writing and was marked by an emerging emphasis on individuality: "The evolution of consciousness through human history is marked by growth in articulate attention to the interior of the individual person as distanced—though not necessarily separated—from the communal structures in which each person is necessarily enveloped" (p. 178). McLuhan was also concerned with the evolutionary process of consciousness and with individuality. Technology is nothing less than an extension of our nervous system, and the result is a profound shift in consciousness: "Electric speed in bringing all social and political functions together in a sudden implosion has heightened human awareness of responsibility to an intense degree" (1994, p. 5). He understood that "fragmented, literate, and visual individualism is not possible in an electrically patterned and imploded society" (p. 51). Concerned less with print and more with emerging electronic technology, he was interested in the changing perspectives, patterns, and mosaics created by different technologies. The structure of print encouraged a separate, specialist, individualistic perspective. By contrast, the electronic media encourage a very different type of perspective.

Teilhard's writing provides a framework for better understanding how evolutionary processes and new social orders might be linked. Teilhard was a Jesuit priest, biologist, and paleontologist, who sought to reconcile Christian theology with the scientific theory of evolution. I will refer to two of his books, *The Phenomenon of Man*, first published in France in 1955, and *The Future of Man* (1964). As a scientist, Teilhard thoroughly understands the power of human action: "Man finds himself overtaken and borne on the whirlwind which his own science has discovered and, as it were, unloosed" (1964, p. 262). He was interested in the consequences of that science of discovery, consequences that he addressed from the perspective of both scientist and priest. Because some considered his ideas unorthodox, the Roman Catholic Church discouraged the reading of his books and forbade him to continue teaching. Not only does he deal with the theory of evolution, but more importantly, he argues that the pressures of our discoveries, inventions, and technologies are leading somewhere, that humankind is evolving. This stance flies in the face of the doctrine that man is created in God's image, suggesting instead that humans are engaged in a crucial process of becoming:

> But now, following the dramatic growth of industry, communications and populations in the course of a single century, we can discern the outline of a formidable event. The hitherto scattered fragments of humanity, being at length brought into close contact, are beginning to interpenetrate to the point of reacting economically and psychically upon each other; with the result, given the fundamental relationship between biological compression and the heightening of consciousness, of an irresistible rise within us and around us of the level of Reflection. (p. 294)

The translator of *The Future of Man*, Norman Denny, explains Teilhard's use of the term *reflection* to mean "the power of conscious thought which distinguishes Man from all other living creatures" (p. 9). In Teilhard's texts, the term *reflection* is also spelled *reflexion*, when the context seems to require it. The use of the term *reflexion* is meant to indicate the way in which consciousness "coils inward upon itself and thus generates new (spiritual) energies and a new form of growth" (p. 9). Contrast McLuhan's point about a growing awareness of interrelatedness with Teilhard's point regarding "an irresistible rise within us and around us of the level of reflection."

Insofar as McLuhan argues for a new awareness of interrelatedness, one must ask, to what end? Can it be that the emergence of new social forms implies no consequence, leads to nowhere in particular? Sir Julian Huxley, in his introduction to *The Phenomenon of Man* (1959), writes that Teilhard:

> ...is rightly and indeed inevitably driven to the conclusion that, since evolutionary phenomena (of course including the phenomenon known as man) are processes, they can never be evaluated or even adequately described solely or mainly in terms of their origins: they must be defined by their direction, their inherent possibilities (including of course also their limitations), and their deducible future trends. (p. 13)

Ong, McLuhan, and others drew on a knowledge of the past, specifically the traditions of preliterate orality, scribe and print cultures, to grasp the present and thereby understand the future. They realized that they were describing evolutionary processes, "of course including the phenomenon known as man." At the very least these three scholars, each in his own way, understood the power of invention to transform the natural world. As Teilhard argues, "one may say that until the coming of Man it was natural selection that set the course of morphogenesis and cerebration, but that after Man it is the power of invention that begins to grasp the evolutionary reins" (1964, p. 293). We are changed by our inventions, and Teilhard discerned a pattern to those changes having to do with the evolution of humankind and the "noosphere." As Huxley explains, Teilhard "coined the term *noosphere* to denote the sphere of mind, as opposed to, or rather superposed on, the biosphere or sphere of life" (1959, p. 13). Ong and McLuhan discern patterns, not only about the cosmos and humankind but about social forms of organization. All three understand that the nature of the social contract is being transformed.

Who we are, how we see the world, and how we form communities, all these and more are affected. Teilhard views the world as an ascending web of reflective interrelatedness; he understands that physical and cultural pressures are forcing humans to become more aware of both themselves and of their relatedness to others. McLuhan writes that "in the electric age we wear all mankind as our skin" (1994, p. 47). It is a typical McLuhan probe that seems to lead nowhere, until this opaque statement is placed next to an equally opaque statement by Teilhard, who writes that "the consciousness of each of us is evolution looking at itself and reflecting" (1959, p. 220). Both statements speak to a new awareness of connection, of the curious effect of being caught in a reflective gaze that cannot be denied.

Ong, McLuhan, and Teilhard each attempt to indicate where that evolution of consciousness is headed; for each of them, the key issue is the intersubjectivity of all communication, the growth of a reflective awareness that transcends individual perspectives. For Ong, "orality-literacy dynamics enter integrally into the modern evolution of consciousness toward both greater interiorization and greater openness" (1982, p. 179). Greater interiorization results in a heightened awareness of one's inner experiences, and greater openness results in a heightened awareness of the experiences of others. Communication technologies have contributed, in unprecedented ways, to a heightened awareness of both inner experience and a greater openness to a world that was, at an earlier time, simply not within one's reach. Ong writes that "both orality and the growth of literacy out of orality are necessary for the evolution of consciousness" (p. 175). If Teilhard is correct that consciousness is evolving, and Ong is correct that orality and literacy were necessary for the evolution of consciousness, does the electronic revolution advance that progress?

Preliterate cultures lacked an adequate concept of what it would mean to be an individual as we use the term today. In a later age, the individual became the measure of all things. Now the concept of individuality has become inherently problematic. As communication technology creates the potential for great isolation, we are also confronted, at the touch of a keyboard, with the images, sounds, and voices of people from all over the world. How can that not have an impact? How can we not be changed? We are accustomed to thinking that an individual can be reflective. It is the autonomous individual who bases his actions on such careful reflection. We are less accustomed to thinking of a larger, collective group as being reflective. We have terms like *groupthink* and *mob-action* but few concepts to capture the possibility of collective sustained reflection. Would we even be able to recognize it?

Transformations

In order to assess Ong, McLuhan, and Teilhard's remarks on self, society, and consciousness, it is necessary to relate their commentary to the world today. The citations provided in the beginning of this chapter are three to four decades old. At that point television was in its infancy; the computer was more an idea than a commodity, and the concept of the Internet would have been mere science fiction. Yet their remarks about the evolution of consciousness capture the implications of the screen. It is imperative to ground not only their specific terms such as *reflexion* or *mosaic* but also their larger claims about the self in the social world. One might argue that such distinctions about the self remain academic, that, regardless of one's theories, humans continue to live and function in communities as they always have. Such an argument would miss the diverse ways in which theory can impact law, public policy, and individual perspectives. The traditions that have shaped American society testify to the power of such ideas.

The screen provides voices and images of a world that was previously beyond our grasp. While we are mesmerized by the content of the screen, we are also influenced by the fundamental features of the medium. Ultimately one is changed not only by the content but also by the subliminal shift in perspective. Though it may be difficult for a viewer to explain, there is the intuition, or perhaps trepidation, that we are simply not grounded as we might have been at an earlier time. The horizon has shifted. There is the inarticulate sense that we experience the world differently. It is not the images and voices of a particular program, the content of the video game, or the digital display that sets one adrift. It is the way in which the screen creates a new phenomenology: we are adrift in a very different experience of space, time, and motion.

A new space conception is created. The physical geography may remain roughly unchanged, even as mediation reconfigures our experience of that geography. It is not that the world grows smaller in a physical sense but that we are acutely aware of the ways in which we are connected politically, economically, and socially. Improvements in transportation, changes in work and play as well as technological innovation, all influence the social domain. One can focus explicitly on the screen without denying that multiple factors have contributed to a transformation of the self.

The social construction of the self is linked to a community's awareness of space and time. Distinctly different views of space and time result in different ways of conceptualizing one's relationship to the larger community. Consider the Buddhist view of reincarnation and the Enlightenment view of rationality. Such concepts also exemplify notions of temporal and spatial forms. Buddhism views reincarnation as a cyclical form that necessitates return. The Enlightenment viewed time as progress and envisioned all aspects, including the self, as part of that move toward the future. The Enlightenment also contributed to a view of the self as an autonomous individual who could fashion life to suit himself. The Buddhist understanding of karma entails a very different concept of the person. Shifts in the collective experience of space and time inevitably influence how the community defines the self in the social world.

A culture's space conception is rarely explicit. It requires education and reflection to recognize that people of different cultures experience perspectives unlike one's own or to imagine spatial and temporal frameworks that are foreign to one's sensibilities. The ways in which a culture thinks of the self are also rarely explicit. The rights afforded the individual and the relationship of that individual to the larger group are expressed through custom, law, and social practice but rarely detailed in explicit ways. Concepts of space and concepts of the self are ways of expressing one's place in the world. How is a particular culture's expression of a space conception related to the ways in which that culture conceptualizes the self?

Art historians, architects, and artists have long recognized that cultural elements can be expressed through spatial forms. The skyscraper of the twentieth century embodies features of modernism that are quite unlike the architecture of eighteenth-century Europe. Not only are cultural elements expressed in spatial form, but such space conceptions influence how a particular culture experiences the self in the social world. Our awareness of what it means to be an individual differs profoundly from earlier ways of thinking about the person. Such shifts can be captured through an analysis of the ways in which a culture conceptualizes space. A new space conception can encourage different ways of relating to the world, which, in turn, may incorporate new ways of thinking about the self.

Humans are grounded in the world, situated in a space and time that embodies cycles of light and dark, rhythms captured by space as well as time. Each of us knows what it means to belong to a particular place, to experience the spatial and temporal forms of our environment. Insofar as the ubiquitous screen alters our experience of space, time, and motion, the altering of social forms may follow. The concept of individualism no longer adequately addresses what it is like to live in a community of the twenty-first century. The challenge is to understand what will replace this quintessential American concept.

American individualism was defined by a historical period that characterized space and time in a vastly different way than we do today. Individualism has been linked to the vastness of an expanding frontier and to the promise of an inexhaustible future. Space existed to be discovered and mastered, just as time existed as a progression of linear moments. This view of space and time embodied a concept of the self as separate from and subtly at odds with the larger community. Our laws and customs are related to that earlier notion of space as a progression to be discovered and mastered. The technology of the book contributed to that view, with its words fixed on the smooth page and its orderly progression of paragraphs and chapters.

The screen contributes to a profoundly different experience of space and time. The promise of an expanding frontier no longer beguiles. The far corners of the world have been surveyed and mapped. The eye does not privilege a singular point of view. The screen teaches one how to see from a variety of angles. Instead of a linear progression, time appears to be relative to our vantage point. The horizon shifts. It seems we are adrift in a world both strange and familiar. There is less confidence in the notion of progress. The autonomy of the individual is called into question because those very landmarks that might secure such a concept are gone, dissolved inadvertently by communication technologies that challenged the possibility of a fixed and singular perspective.

The screen reveals an altered duration; the juxtaposition of image, sound, and motion creates a variable-flex space. One's senses of sight, hearing, and touch are altered, enlarged, enhanced. No longer Euclidian, space floats, morphs, ebbs and flows in ways that are not conceivable in the ordinary world. Yet the screen is now part of our world. Inevitably, the screen influences one's view of everyday life; from an awareness of an urban landscape as mosaic, to an understanding of different ways to frame a close-up, the screen schools one in how to see, feel, and grasp the ordinary world. Without being fully aware of it, we engage a different tempo, sense possible juxtapositions, suspect that sight, sound, and touch are deeply interrelated.

None of this would matter if the self were not phenomenologically grounded in the world. One's existence is grounded in a particular time and place. I know the woods of my backyard, the direction of the rising sun, the lengthening light that comes with spring. None of that is changed. The geography assures me that New York City is east of where I live. Yet something is changed. The geography shifts; the woods, sunrise, and spring do not suffice to explain a fundamental difference. The views of the transcendentalists do not help. The natural world will not reveal what has been altered. The postmodern assumption that the self is merely a multiplicity of social roles does not provide a satisfying answer. The self cannot fully be explained by social circumstances. Nor can postmodernism explain the way in which the screen alters the relationship of self and other.

The transcendentalists advocated listening to a different drummer. Today we do not hear a different drummer so much as viscerally live an altered duration. Individual autonomy requires a separate and singular perspective. Today we are inundated with many perspectives. The screen forces us to be aware of the complex way we are related both to others and to ourselves. It is not the close-up images of distant parts of the world that do this. It is not the instantaneous access to voices and images. More fundamental than that, it is the altering of the human sensorium: the ways in which visual, auditory, and kinesthetic elements are tightly interwoven and then played out in slow and fast motion.

The eye and ear are linked kinesthetically. It is possible to see, feel, and hear the world as never before. That altered duration defines an aesthetic that is characterized by an awareness of boundaries, permeability, and process. What is a variable-flex space but a space-conception defined by duration? A heightened awareness of duration inevitably emphasizes process; boundaries, whether of nation-states or of academic disciplines, become permeable, signifying nodes in a relational network rather than autonomous entities. Just as the technology of print influenced the ways people thought and communicated, the screen, in turn, is influencing awareness of a very different sort. Print contributed to the individualistic perspective. The screen contributes to a definition of the self that cannot be characterized as either individualistic or collectivistic. That dichotomy no longer suffices.

Neither individualism nor collectivism can explain what is occurring. It is not the case that East will meet West or that the West will gradually adopt the perspectives of the East. It is not a contest of cultural will where one side must usurp the other. The question is aptly spelled out by Tu Weiming in a book titled *A Confucian Perspective on Human Rights* (1996). He asks, "Should we understand the self as an isolated individual or as a centre of relationships?" (p. 26). Individualistic and collectivistic elements are evident in the social practices of both East and West. Eastern traditions have long understood that each element inevitably internalizes the other.

There will soon come a point when it is no longer feasible to think in terms of collectivism or individualism. Neither the sanctity of the group nor the autonomy of the individual will prevail. Imagine the relationships that might emerge when a community can no longer be defined either by individualism as we know it or by collectivism. That dichotomy, whether false or not, provided a framework for thinking about the relationship of self and other. We are in need of a new framework, a way of conceptualizing what occurs when that dichotomy no longer serves as a reference point for understanding the social domain.

Comparing the consequences of the book with the consequences of the screen can reveal the shape of things to come. Reading allows for a kind of sensory retreat. The eye scans the smooth page in silence, the inner voice translating phonemes into meaningful units. Books encourage a type of reflection that may, in fact, be peculiar to that activity. As has been pointed out by McLuhan and others, the alphabet encodes the world, and the reader is influenced by that code in the process of translation. Though books provide a doorway to a larger world, it is also apparent that much is lost in the process. The screen challenges the agenda of the book. The eye is no longer privileged. The screen does not encourage the translator's retreat. Awash with image, sound, and motion, the viewer is absorbed in a kinesthetic process that links sight and sound. It is not the eye or the ear, but all the senses together that are extended, altered. The communication matrix is privileged.

There is some intriguing evidence that this highly mediated environment is having an impact on intelligence scores. In an article titled "Rising Scores on Intelligence Tests" published in *American Scientist* (1997), author Ulric Neisser reports that standardized test scores are going up. The largest increases appear not on "such content-related tests as vocabulary, arithmetic or general information," but on tests such as the *Raven's Progressive Matrices*, which measures one's ability to analyze visual forms while keeping track of related hierarchies and goals. "These gains are not limited to the WAIS [Wechsler Adult Intelligence Scale], to adults or to the United States....Data from a dozen other countries show similar trends" (p. 442). Though there could feasibly be many different reasons for such gains, one possible explanation is the prevalence of a "visual and technological" environment:

> Perhaps the most striking 20th-century change in the human intellectual environment has come from the increase in exposure to many types of visual media. From pictures on the wall to movies to television to video games to computers, each successive generation has been exposed to far richer optical displays than the one before. (p. 446)

In *Frames of the Mind: The Theory of Multiple Intelligences* (1983), Howard Gardner argues that there are different types of intelligence. If there is a visual form of intelligence that is linked to abstract reasoning, as studies in geometry might suggest, then an enriched visual environment certainly might influence that type of intelligence.

Though there is not yet empirical evidence to support this hypothesis, "it is possible...that exposure to complex visual media has produced genuine increases in a significant form of intelligence" (Neisser, 1997, p. 447), exactly as Arnheim would have argued. Meyrowitz also suggests "that electronic media may aid the development of other forms of 'intelligence' that we do not yet even know how to name or measure" (1985, p. 326). Such an idea challenges the simple assumption that reading will increase intelligence, while watching the screen will decrease intelligence. Our language reveals the inherent bias. The terms *watching* or *viewing*, as in viewing the screen, may have been appropriate in the early days of television many decades ago, but such terms no longer capture our current involvement with the screen.

The assumption that watching a screen is a more passive activity than reading a book reveals the extent to which many have misunderstood the screen. The screen and the book are, as Carpenter points out, different languages. To better understand that difference, one must investigate the sensory bias of each technology. Compare the process of translating a line of phonemes into meaningful units with the process of watching three people discuss a political topic on the screen. These very different experiences cannot be adequately judged in terms of more or less.

The consequences of each technology are also very different. The screen encourages a way of being in the world, a sense of connection, an experience of simultaneity, that stands in striking opposition to the traditions of print. The child growing up with the screen lives in a new era. Not only does she think in a three-dimensional space as easily as in words, she also implicitly recognizes a way of being with others that was not evident within the traditional individualistic/collectivistic dichotomy.

One is grounded within a horizon that indicates a particular time and place, a horizon that is first and foremost phenomenological. Change the nature of that horizon, alter the phenomenology, introduce new ways of perceiving the world, and awareness of self will be changed as well. Human nature requires that one know oneself through the process of being present in a world. The screen changes how we are present to ourselves and to others. We are not disembodied by the screen; we are re-embodied. Our sensory powers are extended and altered. The self is reconfigured, and in that process awareness of self and other is transformed.

I do not suggest the radical emergence of a new and unprecedented form any more than the emergence of individualism at an earlier time was unprecedented. The dictum that there is nothing new under the sun still holds. The gradual focus on the autonomy of the person, arising in Western Europe during the Enlightenment, was a response to specific conditions. Current shifts in the conceptualization of the self in the social world also arise in response to specific conditions. Change is the only constant. Just as the idea of individualism might have been unthinkable for a preliterate oral community of an earlier era, future ways of conceptualizing the self might be equally unthinkable for us.

To suggest that the self is transformed is not to advocate a radical break with either the past or future. The past is a landmark that the future claims and reconfigures. Inevitably, the process of change results in new forms. The future draws on the customs and traditions of an earlier era in order to create new maps. The self embodies its own geography. Thoreau offers one kind of map. Jung, in turn, offers another version. The Buddhist provides a map that alludes to a vastly different sort of geography. To suggest that the self is being transformed means that a new map is being created. We know only that there will always be maps, that the terrain can be surveyed in diverse ways, that the landmarks will inevitably shift, and that, from time to time, humans manage to lose their way. That is the nature of our predicament.

The printing press led to a transformation in how people thought about ideas. Though it might appear to be a simple example in retrospect, it was actually a complex process that resulted in new ways of thinking about not only ideas, but persons, economics, and medicine. In that earlier era people might not have been aware of the broader implications of print. In turn, we may be less aware of the implications of the screen. As Katsh argues, "it is not surprising that our generation is as oblivious to the cultural, institutional, and psychological ramifications of the electronic media as early printers were to the consequences of printing" (1995, p. 145). The way in which the screen maps ideas is very different from the way in which the book maps ideas. These two technologies, book and screen, encourage vastly different views of the world. The book encouraged the belief that it might be possible to give a detailed and complete account of the world; the screen encourages the awareness that such an account can never be complete. The systematic classification that was possible with the book gives way to an awareness of varied perspectives. The traditions of print are challenged by a screen that graphically portrays motion, sound, and image—forms that could only be implied through print.

A New Perspective

In an earlier era, the knowable horizon was limited by how far one could walk. Then various forms of transportation extended people's awareness of what was beyond the next horizon. Now the screen alters our awareness of geography by revealing a series of possible horizons. The way in which a car transports one is very different from the way in which the screen can transport one. As explained in Chapter Three, the screen is a portal, allowing for a special kind of transport. The senses are altered and extended, even as the body remains fixed in one location. Awareness of a particular place and time is juxtaposed with the mediated experience of a variety of horizons. Now the self transcends the physical form to include a series of possible locations, as points on a continuum or as a relational network. We are more aware of space and time as duration than as fixed points on a map. The geography appears to be in flux. The old map will not explain this new awareness.

We are grounded in the world; we exist within a particular horizon, yet our ways of gathering and sharing information contribute to an experience of multiple horizons. This contradiction, that we are and are not grounded within a particular horizon, implies that we are not merely in the world but that the world is in us. This was also Merleau-Ponty's point: that one's phenomenological awareness emerges from the complex intertwining of self and world. The suggestion that the world is in us could merely imply that our physiology shares the same chemical structures as the rest of the world or that our biological rhythms respond to the shifting patterns of light and dark. Though each of these points is correct, the world is in us in yet another way.

Humans have a profound influence in the world. We invent new technologies, new languages, new political and social structures. The world, in turn, has an equally profound influence on us. Such influence extends far beyond the physiological effect of sunlight hitting the retina.

As a consequence of the screen's influence, we internalize a new experience of space and time. The screen alters the communication matrix, not by changing the structure of the eye or reinventing the physiology of the ear but by providing a different sensory experience. The power of the screen resides in its ability to influence what McLuhan refers to as the ratio of the senses. McLuhan suggests that technologies have the power to isolate the senses. In *The Gutenberg Galaxy*, he writes:

> It is necessary to understand the power and thrust of technologies to isolate the senses and thus to hypnotize society. The formula for hypnosis is "one sense at a time." And new technology possesses the power to hypnotize because it isolates the senses. (p. 272)

Print isolated the senses in the way that McLuhan describes. Print also encouraged a type of reflection that contributed to a view of the self as separate from the larger group. The screen does not isolate the senses; just the reverse is true. The screen links visual, auditory, and kinesthetic elements, altering one's experience of time, space, and duration in the process. One is both physically grounded within a particular horizon and yet not grounded. The consequence is a deep contradiction. With the advent of the screen, it is possible to experience a duration that is nothing like one's ordinary experience. To the extent that such an altered duration is internalized, the world is in us—is influencing us—in a very different way.

When the self is conceptualized as a series of possible points in a network, the autonomous individual can no longer be the measure of all things. American individualism does not suffice to capture this way of being in the world. Individualism favored the autonomy of a singular perspective. The current environment challenges that autonomy. Nor is this way of thinking about the self in the social world simply more collectivistic. What emerges is neither an individualistic society nor a collectivistic society.

A relational network contrasts the singular view of one point on a continuum and the larger pattern. A variable-flex space is characterized by movement, duration. The self in the social world can also be characterized by duration. The individual exists in a variable-flex space. The points on the continuum are important insofar as they reveal the emerging pattern. Do not think just of the pattern (the collective) or just the point on a continuum (the individual). Think instead about the relationship between the single point and the larger pattern. This view of the self in the social world focuses on the shifting interaction of figure and ground.

Neither the transcendentalists nor the postmodernists adequately capture the current geography. Both transcendentalism and postmodernism hold fast to a concept of the self that properly belongs to the nineteenth and twentieth centuries, a concept of the self that favors the autonomy of the individual. American transcendentalism suggested a retreat from society in order to discern the truth of one's soul. The postmodernists allude to the difficulty of grounding one's answers in anything more than the varied perspectives of multiple identities. Yet even the varied perspectives of multiple identities suggest a framework that continues to conceptualize the self in the social world first from one perspective, then from another. There is something more than the truth of the solitary soul. There is something prior to the multiple truths of varied perspectives. The profound challenge of the twenty-first century is to recognize how the self in the social world is no longer grounded in the concept of individual autonomy that emerged with the Enlightenment. The most straightforward way to address this challenge is to detail the starting point for both transcendentalism and postmodernism in order to see the alternative.

As detailed in Chapter Five, American individualism necessarily implies a distinction between private and public life, between inner and outer experience. In order to exist as an individual, one must be separate in some way from the constraints of the social milieu. Such separate experiences are understood to be private and can be viewed as part of one's inner life. The approach of American transcendentalism was to favor the private domain, to listen to one's soul in order to discern the true path. By contrast, postmodernism favors the public domain, defining the individual in terms of those varied social identities that constitute the person. Regardless of how different these two approaches are, they each acknowledge the fundamental distinction of private and public by privileging one over the other. Each of these frameworks requires a choice: either the self is discovered through private reflection or constructed through social encounters. The same choice surfaces when considering whether to define a particular society as individualist or collectivistic.

The difficulty is that the fundamental distinctions of public and private, inner and outer, no longer capture what is essential about the self. Though the concept of individualism once shaped American discourse, that is no longer the case. Neither transcendentalism nor postmodernism will serve as a guide in this new terrain, because neither of these approaches provides an appropriate response to our highly mediated environment. Nevertheless, both transcendentalism and postmodernism have something of value to offer. Solitary reflection can serve as an excellent guide for understanding one's aims and purposes. In turn, individuals can be defined, in part, by their social roles. Yet each of these views of the self gives up too much. The task is to discern the new form that emerges when these different perspectives are folded together.

I do not question that the self exists as a series of possible perspectives, nor do I doubt that the self is defined through a reflective awareness that requires some sort of separation from the larger community. These two views are not mutually exclusive. The book increased the possibilities for creating separation from the larger community by enhancing what Ong referred to as *interiority* or awareness of one's inner life. The screen alters how we think of and experience multiple perspectives by creating a duration that is distinctly different from and yet related to our ordinary experience of the world. Goethe's genius was to understand that it is not the uniqueness of the individual that matters but the unique way in which he chooses to interact with his social milieu. Both the book and the screen provide opportunities for interactions with various communities. Goethe also argued, as detailed in Chapter Five, that "the very specificity of an individuality is tied to the unrepeatable uniqueness of a spatial-temporal point" (Weintraub, 1978, p. 352). For Goethe, the individual stands within a particular horizon and is defined by that singular space and specific moment in a way that is incommensurable with any other place or time. Here is the pivotal distinction between an eighteenth-century view of the self and a twenty-first-century view of the self. It is much more difficult today to place one's faith in the uniqueness of a spatial-temporal point as the defining feature of the individual. We sense viscerally the possibility of viewing any spatial-temporal point from a variety of perspectives.

When transcendentalism and postmodernism are folded together, a different view of the self emerges, more akin to systems theory and less related to a Newtonian world view. Gregory Bateson detailed one possible answer in *Steps to an Ecology of Mind* (1972). For Bateson, the individual is part of a larger system. Understanding the self as part of a system is especially difficult because

> The total self-corrective unit which processes information, or, as I say, "thinks" and "acts" and "decides," is a *system* whose boundaries do not at all coincide with the boundaries either of the body or of what is popularly called the "self" or "consciousness"; and it is important to notice that there are *multiple* difference between the thinking system and the "self" as popularly conceived. (p. 319)

Within this systems theory approach, the self does not exist as a unique spatial-temporal point. From the perspective of systems theory, individual autonomy is less significant than the way in which the self engages the network. This systems view of the self is compatible with McLuhan's point that communication technologies serve as extensions of the human nervous system. It is possible that McLuhan was talking about the self as an ecological system. If the self does not "coincide with the boundaries either of the body or of what is popularly called the 'self,'" it is possible to think about the ways in which the human nervous system may be interwoven with the larger world. Perhaps McLuhan was right in that communication technologies can, in fact, serve as extensions of consciousness.

American culture is encumbered by the traditions of the past and confused by the implications of the present. The concept of individualism emerged in response to a particular time and place. Our laws and traditions are based on a unique concept of the self that navigates the twin concerns of autonomy and community. This view of the self once helped a new nation to define the inalienable rights and responsibilities of the person.

We now live in a very different era. The social milieu of the twenty-first century is quite different from that eighteenth-century perspective. The individualistic perspective and the systems theory approach illustrate that difference; the singular viewpoint is disavowed in favor of a reciprocal and shifting vantage point. Though the language of individualism may be embodied in our laws and customs, our everyday experience suggests something else. In an article titled "Ecology and Some of Its Future" (2002), Ong writes that "interconnectedness is the mark of our age, at least as apprehended by the human sensibility" (p. 11). Our heritage encourages a way of thinking about the self that privileges the autonomous individual. Yet we live with the awareness of a duration that intensifies the possibility of connection. Our ways of thinking about the self and our ways of experiencing the self are at odds.

The American experiment with democracy identified the tensions that exist between the rights of the individual and the needs of the community. Despite that heritage, or perhaps because of that heritage, the current age displays very different tensions. The future demands that we address the contradictions of our past. We cannot claim, as the colonies did, to be a country without traditions. America is constrained by that individualistic concept of the self. The current environment is caught between that concept of individualism and a concept of the self in the social world that embodies a new form. To the extent that the altered duration of the omnipresent screen is internalized, we experience a space conception that leads inevitably to an experience of the self that is at odds with individualism.

The challenge is to understand what it might mean to exist as an individual in such a system. Where is the self located in such a network? What grounds the self? Can the self exhibit continuity over time? The question "where is the self located?" implies the possibility of a unique spatial temporal point, a fixed horizon, a space conception that is at odds with the current environment. To examine something from a systems theory point of view is to explore the phenomenon from multiple vantage points, hence the postmodern inclination to conceptualize the self as varied social roles.

The postmodern response privileges the system, construing the self as a shifting element in a social network. The difficulty is that the self is not merely subsumed by the system; too much is lost with that approach. Both the system and entities within that system can be understood as reciprocal elements. The continuity of the self over time is embodied in a way of being that is defined by the lived body. There is an ontology, inherent in physiological and phenomenological processes, that constitutes a matrix that prepares the way for the emergence of the larger system.

To conceptualize the self as the construal of various social identities is to leave something unaccounted for, to overlook a fundamental element. As explained by Merleau-Ponty, we are grounded in the world and yet are of the world. That decentering makes it possible to take a perspective, to see things from one vantage point rather than another, to recognize what it means to live situated within a particular horizon. Without such a phenomenology, it would not be possible for the screen to alter one's perception of that horizon. It is ironic that postmodernists position the self as merely a series of relationships. One must be situated in the world in the way Merleau-Ponty indicates in order to have a vantage point from which to reflect on that series of networks.

I am more than various nodes in the larger network of social connections. The system does not subsume me because, in addition to participating in the system, I also have a perspective on the system. Understanding the self as an element in a system does not negate the autonomy of the person. It is possible to alter my relationship to that system and, in doing so, to alter the system in unpredictable ways. The self is, and is not, the system. The self is, and is not, the autonomous individual seeking a path through the world. To view the self as an individual misleads. To view the self as a collective misleads. The alternative is to view the self and the larger system as reciprocal elements circumscribed by connection. This alternative requires that one understand both the nature of connection and the self in a way that is at odds with previous notions of individualism.

The varied connections of the system entail more than social relationships. There are biochemical, sociological, ecological, psychological, and intellectual ways of existing in relationship, to name a few.

Science teaches us that we are not separate entities. Far from it, we breathe the air from each other's lungs, react to cycles of light and dark, share viruses as well as ideas, respond to social nuance with biochemical changes, and react to forms of communication that will never be captured in language. We are perhaps more conscious in the twenty-first century of the ways in which we are connected to others, to ourselves, and to the physical world than would have been possible in an earlier era. What was once a mystical intuition now becomes a scientific reality. We recognize that the self is defined by boundaries that are both permeable and fluid. Living entities do not exist as separate creatures. The self is neither separate from nor entirely defined by the social domain.

As part of the emergent system, the self shares the features of, and is defined by, the constraints of that system. The system is not fixed, it evolves; it cannot be mapped as an exact structure; it is predictable to a point but ultimately escapes our best attempts at prediction. In such a system it is difficult to identify where the person ends and the rest of the environment begins. The system exhibits conscious and unconscious properties, is reflective and, as a whole, is seemingly incapable of reflection. The self is and is not the system, participates in the system, and yet can partially transcend the system to take a perspective on elements of that system. It is difficult to grasp the nature of that reciprocal relationship between the self and system. We have less difficulty positing the inalienable rights of the individual than in understanding the ways in which the individual contributes to the larger whole. In Western culture, the individual is understood as either standing apart or as part of the social milieu but not as a blend of the two. It is difficult to imagine how such a blend would allow for the autonomy of the person. And yet such a system depends on that autonomy and cannot function without it.

The system thinks. It is aware. That is possible because the individual who stands in a reciprocal relation to the system is aware. The system does not exemplify group-think or mob action. Something more is possible, though our language does not offer readily available terms to describe what that would be.

Teilhard describes it as the *noosphere*, as *reflexion*. In order to understand what he might have meant, we must give up nothing less than the foundations of our culture; we must question what it means to be an individual in the twenty-first century. In the past, American culture conceptualized experiences as either individual and personal or social and shared. That distinction is reflected in everything from academic disciplines to epistemology. Now we must grasp the reciprocal ways in which individual and social perspectives are related. We cannot choose one over the other. The world as we now know it will not allow for that option. It is not enough to suggest that the person is, in some way, both individual and social; we must recognize what unifies these domains.

There are crucial elements of the individualistic perspective that become problematic for a systems theory approach. As discussed in Chapter Five, individualism distinguishes between inner and outer experience, private and public life. The current communication environment contributes to an awareness of self and other that undermines such distinctions. The figure-ground has shifted. Where once the self was defined by the distinction of inner and outer, private and public, that dichotomy no longer captures what is essential.

If individualism no longer serves as an explanation of the self in the social world, how do we think about these categories? How could it be that the fundamental distinction of private and public no longer captures the relationship of self and other? It is possible that the same historical circumstances that contributed to the emergence of individualism also contributed to such categories. There may be other historical periods and other cultures that do not share this way of conceptualizing the person. Carpenter details the way in which the Inuit characterize all experience as outer (1966). For this culture there is no inner experience. The challenge is to grasp the transition from one way of thinking about the self to another. What are the defining features of a self that does not distinguish between inner and outer experience?

Topics once considered quite personal are now freely discussed in public discourse. From the behavior of presidents to the lives of ordinary citizens, we find ourselves entangled in conversations that would have been taboo in an earlier era. Some argue that the media's desire for the sensational story motivates this sort of trespass into the private domain. That assessment conceals a deeper issue. The media reveal a shift in shared attitudes regarding what topics are appropriate for public discourse, but the media are not responsible for creating that broader shift in attitudes. The boundary between private and public life is dissolving. Insofar as the terms *private* and *public* share a family resemblance with the terms *inner* and *outer*, it is likely that shared attitudes regarding inner and outer experience are shifting as well.

With individualism, the autonomy of the person was foregrounded. When the self is understood from the perspective of systems theory, connection, rather than autonomy, is foregrounded. With this figure-ground shift, the autonomy of the individual does not disappear: it moves to the background and in doing so supports the foreground. Rather than emphasizing the distinct and individualized elements of the self, the emphasis shifts to an awareness of the various ways in which we are connected or disconnected as the case may be. Insofar as connection is privileged, distinctions of inner and outer experience become less significant, as does the distinction between private and public life. The emphasis shifts to understanding the deeper implications of relationship. And that is why the current environment encourages a concern with isolation and alienation.

In an earlier century, the emphasis was on freedom. The task then was to secure the inalienable rights of the individual. The task now is to discern the varied ways we stand in relationship to ourselves and others. The tensions between self and community that once defined the individualist perspective cannot help us understand the current landscape. What might it mean that the boundary between inner and outer experience dissolves? If connection is the defining motif of this age, why then do we so often feel disconnected?

Familiar landmarks no longer serve as a guide. We are in need of a vastly different way of thinking about the self that transcends the categories of an earlier age. Awareness of one's relationship to a larger system appears to threaten the autonomy of the individual. It is difficult to replace the individualist perspective with something else. The highly mediated environment makes us aware of the way in which everything is in flux, of the way in which everything is connected. And, by contrast, it appears we are more disconnected than ever, caught in the reflections of a screen that portrays the world at a distance, visible and yet beyond our grasp. We are not immune to the consequences of the screen; we are immersed, not disembodied but re-embodied.

Those who are concerned with the isolating effects of technology are implicitly attuned to the ways in which the screen alters our awareness of what it means to be connected. To be isolated is to be isolated from something. What might this highly mediated environment isolate us from? Bateson argues that "it is...of prime importance to have a conceptual system which will force us to see the 'message' (e.g., the art object) as *both* itself internally patterned *and* itself a part of a larger patterned universe" (1972, p. 132). Substitute the word *self* for the word *message* in the above quote. To paraphrase, such a conceptual system forces us to see the self as both internally patterned and as a part of a larger patterned universe. The variable-flex space of the screen awakens the intuition of connection, process, and pattern. The arduous task is to grasp the ways in which the self is part of that larger patterned universe. This task is especially difficult for a culture that has privileged an individualistic perspective.

The horizon shifts; our experience of time and space is altered; a new space conception challenges the culture. Individualism has run its course. We must find new ways to think about the self that are compatible with our experiences. The distinctions of inner and outer, private and public, do not adequately explain the ways in which self and system might be interrelated. A different framework is needed. Such a framework is evident in our ordinary ways of speaking about the world. Return to the perceptual gestalt to discern what is privileged, what is foregrounded. The concept of individualism privileged private, inner experience. The system foregrounds a heightened awareness of connection, of the ways in which one is related to other people, places, events, ideas.

Current fascinations with connection are evident in political and social discourse. Consider the political interest in globalization, questions about the nature of the nation-state. Note the concerns regarding our growing isolation, the fear that communication technologies contribute to a deepening sense of alienation. The category of private experience may not help one understand one's "place" as part of a larger interrelated system that is in flux. We live now in a variable-flex space. We seek to understand how we are connected in that variable-flex space.

The system is more than the roles one plays in the social domain. To understand the way in which the system is in flux, one need only to think of emergent patterns evident in chaotic forms, of the ways in which complexity contributes to change. To understand how the system is grounded, one need only to think of that phenomenological awareness of a particular horizon, the basic recognition that one's physical presence in the world offers the first and most important connection. As Ong noted, "The body is a frontier between myself and everything else" (1982, p. 73). That is the foundation.

References

Altheide, D. L. (1995). *An ecology of communication: Cultural formats of control.* New York: Walter de Gruyter.

Arnheim, R. (1954). *Art and visual perception* (4th ed.). Berkeley, CA: University of California Press.

———— (1957). *Film as art* (4th ed.). Berkeley, CA: University of California Press.

———— (1965). Visual thinking. In G. Kepes (Ed.), *Education of vision* (pp. 1–15). New York: George Braziller.

———— (1971). *Visual thinking.* Berkeley, CA: University of California Press.

Bateson, G. (1972). *Steps to an ecology of mind.* New York : Ballantine.

Baudrillard, J. (1988) The masses: The implosion of the social in the media. In M. Poster (Ed.), *Selected writings.* (pp. 207–219). Stanford, CA : Stanford University Press.

Bellah, R. N., et al. (1985) *Habits of the heart : Individualism and commitment in American life.* New York : Harper & Row.

Berger, P. L. & Luckmann, T. (1967). *The social construction of reality : A treatise in the sociology of knowledge.* Garden City, NY : Anchor Books.

Bianculli, D. (1992). *Teleliteracy: Taking television seriously.* New York: Continuum.

Bower, B. (2001). Joined at the senses: Perception may feast on a sensory stew, not a five-sense buffet. *Science News, 160,* 204–205.

Carey, J.W. (1992). *Communication as culture: Essays on media and society.* New York: Routledge

Carpenter, E. (1960). The new languages. In M. McLuhan & E. Carpenter (Eds.), *Explorations in communication: An anthology* (pp. 162–179). Boston: Beacon.

———— (1966). Image making in Arctic art. In G. Kepes (Ed.), *Sign, image, symbol* (pp. 206–225). New York: George Braziller.

Carpenter, E. & McLuhan, M. (1960). Acoustic space. In M. McLuhan & E. Carpenter (Eds.), *Explorations in communication: An anthology* (pp. 65–70). Boston: Beacon.

Chardin, T. de. (1959). *The phenomenon of man* (B. Wall, Trans.). New York : Harper & Row.

———— (1964). *The future of man* (N. Denny, Trans.). New York : Harper & Row.

———— (1966). *The vision of the past* (J. M. Cohen, Trans.). New York : Harper & Row. (Original work published 1957)

Corwin, N. (1989). "Give us 22 minutes. . . ." In W. K. Agee, P. H. Ault, & E. Emery (Eds.), *Maincurrents in mass communications* (pp. 26–30). New York: Harper & Row.

Croteau, D. & Hoynes, W. (2000). *Media/society: industries, images, and audience.* (2nd ed.). Thousand Oaks, CA: Pine Forge.

Czitrom, D. J. (1982). *Media and the American mind : From Morse to McLuhan.* Chapel Hill : University of North Carolina Press.

Dance, F. E. X. (1979). Acoustic trigger to conceptualization : A hypothesis concerning the role of the spoken word in the development of higher mental processes. *Health Communications and Informatics, 5,* 203–213.

Denzin, N. K. (1991). *Images of postmodern society : Social theory and contemporary cinema.* London: Sage Publications.

Denzin, N.K. & Lincoln, Y. S. (Eds.). (2000). *Handbook of qualitative research* (2nd ed.). London : Sage Publications.

Dillon, M. C. (1988). *Merleau-Ponty's ontology.* Bloomington, IN : Indiana University Press.

Dorfles, G. (1965). The role of motion in our visual habits and artistic creation. In G. Kepes (Ed.), *The nature and art of motion (pp. 41-50).* New York : George Braziller.

Drucker, J. (1995). *The alphabetic labyrinth: The letters in history and imagination.* London: Thames and Hudson, Ltd.

Eisenstein, E. L. (1983). *The printing revolution in early modern Europe.* New York: Cambridge University Press.

Emerson, R. W. (1975). *Self reliance.* St. Helena: Illuminations Press. (Original work published 1841)

Fiske, J. (1987). *Television culture: Popular pleasures and politics* (Studies in Communication). New York: Routledge.

Friedhoff, R. & Benzon, W. (1989). Visual experiments. In E. M. Pavese (Ed.), *Visualization: The second computer revolution* (pp. 132–171). New York: Harry N. Adams.

———— (1989). Computer graphics. In E. M. Pavese (Ed.), *Visualization: The second computer revolution* (pp. 82–131). New York: Harry N. Adams.

Gardner, H. (1983). *Frames of mind: The theory of multiple intelligences.* New York: Basic Books.

Gergen, K. (1991). *The saturated self: Dilemmas of identity in contemporary life.* New York: Basic.

Giedion, S. (1960). Space conception in prehistoric art. In M. McLuhan & E. Carpenter (Eds.), *Explorations in communication: An anthology* (pp. 71–89). Boston: Beacon.

———— (1966). Symbolic expression in prehistory and in the first high civilizations. In G. Kepes (Ed.), *Sign, image, symbol* (78-91). New York: George Braziller.

Goody, J. (1977). *The domestication of the savage mind.* New York: Cambridge University Press.

———— (1986). *The logic of writing and the organization of society.* New York: Cambridge University Press.

——— (1987). *The interface between the written and the oral*. New York: Cambridge University Press.

——— (1990). The *oriental, the ancient and the primitive: Systems of marriage and the family in the pre-industrial societies of Eurasia*. New York: Cambridge University Press.

——— (1996). *The east in the west*. New York: Cambridge University Press.

———(2000). *The power of the written tradition*. Washington: Smithsonian Institution Press.

Gozzi, R. (1990). *New worlds and a changing American culture*. Columbia, SC: University of South Carolina Press.

———(1999). *Power of metaphor in the age of electronic media*. Cresskill, NJ: Hampton.

Grodal, T. (1997). *Moving pictures: A new theory of film genres, feelings and cognition*. New York: Clarendon.

Grosz, E. (1994). *Volatile bodies: Toward a corporeal feminism*. Bloomington, IN: Indiana University Press.

Hall, E. (1966). *The hidden dimension*. New York: Doubleday.

Havelock, E. A. (1963). *Preface to Plato*. Cambridge, MA: Harvard University Press.

——— (1986). *The muse learns to write: Reflections on orality and literacy from antiquity to the present*. New Haven, CT: Yale University Press.

Hay, J. (1992). Afterword. In Robert C. Allen (Ed.), *Channels of discourse, reassembled*. (pp. 354–387). Chapel Hill, NC: University of North Carolina Press.

Hayles, N. K. (1999). *How we became posthuman: Virtual bodies in cybernetics, literature, and informatics*. Chicago: University of Chicago Press.

Heelan, P. (1983). *Space perception and the philosophy of science*. Berkeley, CA: University of California Press.

Ihde, D. (1986). *Consequences of phenomenology*. Albany, NY: State University of New York Press.

——— (2002). *Bodies in technology*. Minneapolis, MN: University of Minnesota Press.

Jagodzinki, C. M. (1999). *Privacy and print: reading and writing in seventeenth century England*. Charlottesville, VA: University Press of Virginia.

Jamieson, K. H. (1988). *Eloquence in an electronic age. The transformation of political speech making*. New York: Oxford University Press.

——— (1988). *The interplay of influence: Mass media and their politics in news, advertising, politics* (2nd ed.). Belmont, CA: Wadsworth.

Jay, M. (1993). *Downcast eyes: The denigration of vision in twentieth-century French thought*. Berkeley, CA: University of California Press.

Jung, C. (1957). *The Undiscovered Self*. New York: The New American Library.

Kaha, C. W. (1993). Toward a syntax of motion. *Critical Studies in Mass Communication, 10*, 339–348.

——— (1994). Of boundaries and variable-flex space. *Proteus, 11*, 18–20.

———— (1997). A syntax of motion: Time, space, and television. *Cultural Studies, 2,* 59–70.

————(2000). From radio to television: Space, sound, and motion. *Studies in Symbolic Interaction, 23,* 113–124.

Katsh, M. E. (1989). *The electronic media and the transformation of law.* New York: Oxford University Press.

———— (1995). *Law in a digital world.* New York: Oxford University Press.

Lanham, R. A. (1993). *The electronic word: Democracy, technology, and the arts.* Chicago, IL: The University of Chicago Press.

Levinson, P. (1999). *Digital McLuhan: A guide to the information millennium.* New York: Routledge.

Martin, R. W. T. (2001). *The free and open press: The founding of American democratic press liberty, 1640–1800.* New York: New York University Press.

Matthei, H. (1999). Inventing the commercial. In J. Gorham (Ed.), *Mass media* (pp. 154–163). Guilford CT: Dushkin/McGraw-Hill.

McLuhan, M. (1951). *The mechanical bride: Folklore of industrial man.* New York: Vanguard.

———— (1962). *The Gutenberg galaxy: The making of typographic man.* Toronto: University of Toronto Press.

———— (1964). *Understanding media: The extensions of man.* New York: The New American Library.

———— (1994). *Understanding media: The extensions of man* (30th anniversary ed.). Cambridge, MA: MIT Press.

McLuhan, M. & Parker, H. (1968). *Through the vanishing point: Space in poetry and painting.* New York: Harper & Row.

Merleau-Ponty, M. (1962). *Phenomenology of perception* (C. Smith, Trans.). London: Routledge & Kegan Paul.

———— (1964). *The primacy of perception* (M. Edie, Ed.). Evanston, IL: Northwestern University Press.

———— (1964). *Signs* (J. Wild, Ed.). Evanston, IL: Northwestern University Press.

———— (1968). *The visible and the invisible* (J. Wild, Ed.). Evanston, IL: Northwestern University Press.

———— (1973). *The prose of the world* (C. Lefort, Ed.). Evanston, IL: Northwestern University Press.

Meyrowitz, J (1985). *No sense of place: The impact of electronic media on social behavior.* New York: Oxford University Press.

———— (1998). Multiple media literacies. *Journal of communication, 48(1),* 96–108.

Molinaro, M., McLuhan, C. & Toye, W. (Eds.) (1987). *Letters of Marshall McLuhan.* Oxford: Oxford University Press.

Moore, B. (1984). *Privacy: Studies in social and cultural history.* Armonk, NY: M. E. Sharpe, Inc.

Neisser, U. (1997). Rising scores on intelligence tests. *American Scientist, 85,* 440–447.

Neuman, S. B. (1991). *Literacy in the television age.* Norwood, NJ: Ablex.

O'Neill, M. (1989). The power of the press. In W.K. Agee, P. H. Ault, & E. Emery (Eds.), *Maincurrents in mass communications* (pp. 8–14). New York: Harper & Row.

Ong, W. J. (1967). *The presence of the word: Some prolegomena for cultural and religious history.* New Haven, CT: Yale University Press.

——— (1971). *Rhetoric, romance and technology: Studies in the interaction of expression and culture.* Ithaca, NY: Cornell University Press.

———(1977). *Interfaces of the word: Studies in the evolution of consciousness and culture.* Ithaca, NY: Cornell University Press.

———(1982). *Orality and literacy: The technologizing of the word.* New York: Routledge.

———(2002).Ecology and some of its future. In J. Y. Lee & L. Strate (Eds.), *Explorations in Media Ecology, 1 , 5*–12.

Panofsky, E. (1995). Style and medium in motion pictures. In I. Lavin (Ed.), *Three essays on style* (pp. 91–126). Cambridge, MA: MIT Press.

Peters, J. D. (1999). *Speaking into the air: A history of the idea of communication.* Chicago: University of Chicago Press.

Postman, N. (1985). *Amusing ourselves to death: Public discourse in the age of show business.* New York: Viking.

Riesman, D. (1961). *The lonely crowd.* New Haven: Yale University Press. (Original work published 1950)

Schramm, W. (1988). *The story of human communication: Cave painting to microchip.* New York: Harper & Row.

Schutz, A. (1967). *The phenomenology of the social world* (G. Walsh & F. Lehnert Trans.). Evanston IL: Northwestern University Press.

Sennet, R. (1974). *The fall of public man.* New York: W. W. Norton.

Sola Pool, I. de. (1990). *Technologies without boundaries: On telecommunications in a global age.* Cambridge: Harvard University Press.

Strate, L., Jacobson, R. L., & Gibson, S. (Eds.). (2003). *Communication and cyberspace: Social interaction in an electronic environment* (2nd ed.). Cresskill, NJ: Hampton Press.

Taylor, C. (1989). *Sources of the self: The making of modern identity.* Cambridge: Harvard University Press.

Thoreau, H. D. (1983). Walden *and* Civil disobedience (M. Meyer Ed.). New York: W. W. Norton. (Original work published 1854 and 1849)

Tocqueville, A. de. (1956). *Democracy in America* (R. D. Heffner Ed.) New York: Penguin. (Original work published 1835)

Waite, C. K. (2002). The role of the self in the social world. *Studies in Symbolic Interaction, 25,* 215–231.

Weiming, T. (1996). *A Confucian perspective on human rights.* Singapore: Unipress.

Weintraub, K. J. (1978). *The value of the individual.* Chicago: University of Chicago Press.

Whorf, B. L. (1956). *Language, thought, and reality: Selected writings of Benjamin Lee Whorf* (J. B. Carroll, Ed.). Cambridge, MA: MIT Press.

Wiener, P. (1973). *Dictionary of the history of ideas.* New York: Scribner's.

Williams, R. (1974). *Television: Technology and cultural form.* Hanover: Wesleyan University Press.

Wittgenstein, L. (1958). *Philosophical investigations* (3rd ed.). (G. E. M. Anscombe, Trans.) New York: Macmillan.

———— (1980). *Culture and Value* (P. Winch, Trans.) Chicago: University of Chicago Press. (Original work published 1977)

Wober, J. M. (1988). *The uses and abuses of television: A social psychological analysis of the changing screen.* Hillsdale, NJ: Erlbaum.

Youngblood, G. (1970). *Expanded cinema.* New York: E.P. Dutton & Co.

Zelezny, J. D. (1993). *Communications law: Liberties, restraints, and the modern media.* Belmont, CA: Wadsworth.

Zettl, H. (1999). *Sight, sound, motion: Applied media aesthetics.* New York: Wadsworth.

Index

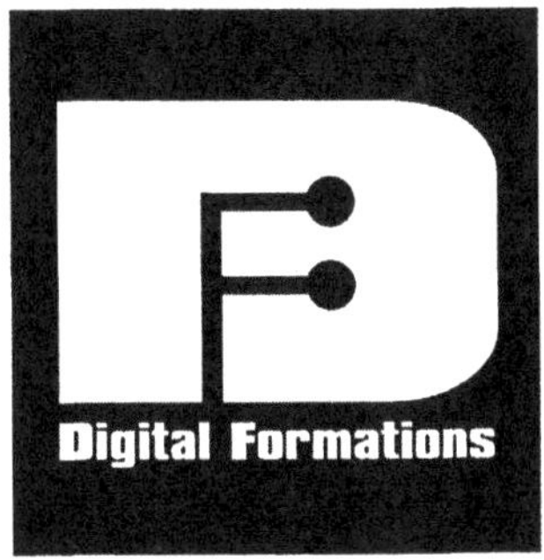

General Editor: Steve Jones

Digital Formations is the new source for critical, well-written books about digital technologies and modern life. Books in this series will break new ground by emphasizing multiple methodological and theoretical approaches to deeply probe the formation and reformation of lived experience as it is refracted through digital interaction. Each volume in *Digital Formations* will push forward our understanding of the intersections—and corresponding implications—between the digital technologies and everyday life. This series will examine broad issues in realms such as digital culture, electronic commerce, law, politics and governance, gender, the Internet, race, art, health and medicine, and education. The series will emphasize critical studies in the context of emergent and existing digital technologies.

For additional information about this series or for the submission of manuscripts, please contact:

Acquisitions Department
Peter Lang Publishing
275 Seventh Avenue 28th Floor
New York, NY 10001

To order other books in this series, please contact our Customer Service Department:

(800) 770-LANG (within the U.S.)
(212) 647-7706 (outside the U.S.)
(212) 647-7707 FAX

or browse online by series:

WWW.PETERLANGUSA.COM